Timber Bridges

Bridges built in timber are enjoying a significant revival, both for pedestrian and light traffic and increasingly for heavier loadings and longer spans. Timber's high strength-to-weight ratio, combined with the ease and speed of construction inherent in the off-site prefabrication methods used, make a timber bridge a suitable option in many different scenarios.

This handbook gives technical guidance on forms, materials, structural design and construction techniques suitable for both small and large timber bridges. Eurocode 5 Part 2 (BS EN 1995-2) for the first time provides an international standard for the construction of timber bridges, removing a potential obstacle for engineers where timber construction for bridges has not – in recent centuries at least – been usual.

Clearly illustrated throughout, this guide explains how to make use of this oldest construction material in a modern context to create sustainable, aesthetically pleasing, practical and durable bridges. Worldwide examples include Tourand Creek Bridge, Canada; Toijala, Finland; Punt la Resgia, Switzerland; Pont de Crest, France; Almorere Pylon Bridge, the Netherlands.

Christopher J. Mettem worked for many years as an engineering consultant at TRADA and has led many research projects in timber structures, composites and connections, conservation work and timber bridges. Through contracts with government agencies he worked in Africa, Central and South America and in the UK with the Highways Agency. He has served on BS committees for structural timber and was a member of the timber Eurocodes drafting teams. He is now the Chairman of the Glued Laminated Timber Association, which he helped to found.

Christopher was one of the three joint authors of *Green Oak in Construction*, which was shortlisted for the RIBA Book Awards, and has also written numerous other TRADA publications.

Timber Bridges was developed and written with the support of the Timber Research and Development Association (TRADA), the centre of excellence for timber expertise and information. Further information on the use of timber in construction and access to the TRADA advisory line service is available at: www.trada.co.uk

Timber Bridges

Christopher J. Mettem

TRADA
TECHNOLOGY

Spon Press
an imprint of Taylor & Francis

First published 2011
by Spon Press
2 Park Square, Milton Park, Abingdon, Oxon OX14 4RN

Simultaneously published in the USA and Canada
by Spon Press
270 Madison Avenue, New York, NY 10016

Spon Press is an imprint of the Taylor & Francis Group, an informa business

© 2011 TRADA Technology Ltd

Typeset in Arial by Wearset Ltd, Boldon, Tyne and Wear
Printed and bound in India by Replika Press, Pvt. Ltd, Sonepat, Haryana

The Timber Research and Development Association (TRADA) – is an internationally recognised centre of excellence on the specification and use of timber and wood products. TRADA is a company limited by guarantee and not-for-profit membership-based organisation. Its name is synonymous with independence and authority.

TRADA Technology, part of the BMTRADA Group, is TRADA's appointed provider for its research and information programmes. It also undertakes a wide range of commercial and training services to the timber and construction industries.

www.trada.co.uk
information@trada.co.uk

British Library Cataloguing in Publication Data
A catalogue record for this book is available from the British Library

Library of Congress Cataloging-in-Publication Data
Mettem, Christopher.
Timber bridges/Christopher Mettem; TRADA Technology Ltd.
p. cm.
Includes bibliographical references and index.
1. Bridges–Design and construction–Handbooks, manuals, etc. 2. Wooden bridges–Design and construction–Handbooks, manuals, etc. 3. Wooden bridges–Specifications–Handbooks, manuals, etc. I. TRADA Technology. II. Title.

TG300.M48 2010
624.2′18–dc22

2009050125

ISBN13: 978-0-415-57796-0 (hbk)

Contents

Foreword   10

1 Benefits of timber bridges   **13**
  1.1   Essentials   13
  1.2   Durability   14
  1.3   Benefits   15
     1.3.1 Sustainability   16
     1.3.2 Forestry benefits   16
     1.3.3 Appearance   18
     1.3.4 Low mass   18
     1.3.5 Low whole-life costs   19
  1.4   Applications   19
  1.5   Modern designs   20
  1.6   Architecture   21
  1.7   Generic timber information   21
  1.8   New materials   21

2 The evolutionary development of the timber bridge   **22**
  2.1   General   22
  2.2   Beams   23
     2.2.1 Under-strutted beams and frames 26
     2.2.2 Railway trestles   26
     2.2.3 North American trestles   27
     2.2.4 Conserving trestles   28
     2.2.5 Contemporary beams   29
     2.2.6 Contemporary under-strutted frames 30
  2.3   Cantilevers   30
     2.3.1 Modern cantilevers   31
  2.4   Suspension bridges   31
     2.4.1 Historic types   31
     2.4.2 Contemporary forms   32
  2.5   Arches   32
     2.5.1 Roman arches   32
     2.5.2 Timber arches in the Far East   33
     2.5.3 English historic arches   34
     2.5.4 Glulam arches   34
     2.5.5 Contemporary arches   35
  2.6   Trusses   35
     2.6.1 North American patented trusses 36
     2.6.2 Contemporary trusses   38
  2.7   Hybrids   39
     2.7.1 Modern hybrids   40
     2.7.2 Moving bridges   40
     2.7.3 Modern versions   41
     2.7.4 Future prospects   41

3 Durability and protection by design   **42**
  3.1   Introduction   42
  3.2   Design working life   43
  3.3   Agents and their effects   44
  3.4   Protection classifications   45
  3.5   Fundamental arrangements   46
  3.6   Biological agents   46
  3.7   Preservation principles and hazard classes   47
  3.8   Extreme hazards   47
  3.9   Hazard classifications for normal bridge components 48
  3.10 Risk evaluation and decisions   48

Contents

3.11 Preservative treatment types 49
3.11.1 Processes 51
3.11.2 Surface treatments 51
3.11.3 Metal-free preservatives 51
3.12 Relating the classifications 51
3.13 Construction protection arrangements 52
3.14 Covered bridges 53
3.15 Cladding and local covers 53
3.15.1 Arrangements of board cladding 53
3.15.2 Louvred cladding 54
3.15.3 Metal covers 55
3.15.4 End grain protection 55
3.15.5 Parapets as protection 55
3.16 Connections 56
3.17 Junctions 56
3.17.1 Deck-to-approach junctions 57
3.18 Decks 57
3.19 Construction protection example: Pont de Merle 58
3.20 Principal timber species in relation to durability design 59

4 Materials **60**
4.1 Introduction: the generic range 60
4.1.1 Steelwork 61
4.2 The main options 62
4.3 Solid timber 62
4.3.1 Dimensions 63
4.3.2 Conformity 65
4.3.3 Strength classes 65
4.3.4 Natural durability and treatability 66
4.4 Glulam 67
4.4.1 Glulam species 68
4.4.2 Durability categories for glulam 69
4.4.3 Standards for glulam 69
4.4.4 Glulam strength classes 69
4.5 Mechanically laminated timber 70
4.6 Structural timber composites 71
4.6.1 Laminated Veneer Lumber (LVL) 71
4.6.2 Innovative STCs 72
4.6.3 Structural plywood 72
4.7 Innovations 72
4.8 Metal fasteners, connectors and corrosion protection 73
4.9 Structural adhesives 74
4.9.1 New structural adhesives 75
4.10 Cladding and shielding timbers 75
4.10.1 Heat-treated cladding 76
4.11 Forest certification and chain of custody 77

5 Concept design **78**
5.1 Introduction 78
5.2 Applications 78
5.2.1 Feasibility 78
5.3 The structural essentials 79
5.3.1 Arrangement decisions 81
5.3.2 Prefabrication and erection 81
5.4 Architecture 83
5.4.1 Form 83
5.4.2 Aesthetics 83
5.4.3 Material textures, finishes and details 84
5.5 Terrain 85
5.6 Geology and site access 85

5.7   Location, purpose and functional geometry   86
5.7.1 Parapets   87
5.7.2 Enclosed footbridges   87
5.8   Rapid construction for a railway: Passerelle de Vaires-sur-Marne   88
5.9   Choices of structural form   88
5.9.1 Deck elevations   89
5.10 Beams   89
5.10.1 Two-way spanning slabs   92
5.11 Cable-stayed, suspension and tension-ribbon bridges   93
5.11.1 Tension-ribbons   94
5.12 Arches   95
5.12.1 General arrangements   95
5.12.2 Arch geometry   96
5.12.3 Bracing   96
5.12.4 Cross-sections   97
5.13 Trusses   97
5.14 Summary: the main aspects of a successful timber bridge   99
5.14.1 Principal components   100
5.14.2 Deck design   100
5.14.3 Other functions   100
5.15 Obtaining outline approval for the design   101

6 Decks and parapets   102
6.1   Introduction   102
6.2   General functions   103
6.2.1 Protection   103
6.3   Solid timber decks   103
6.3.1 Laminated veneer lumber   105
6.4   Laminated decks   106
6.5   Cross laminated plank decks   106
6.6   Nail laminated decks   107
6.7   Stressed laminated deck design concepts   108
6.8   Stressed laminated glulam decks   111
6.9   Timber–concrete composite decks   112
6.10 Innovative all-timber decks   113
6.11 Wearing surfaces for vehicular bridges   113
6.11.1 Sealing   113
6.12 Parapets and handrails   114
6.12.1 Cantilevered parapets   115
6.12.2 Demountable parapets   116
6.12.3 Handrails   116
6.13 Ramps, steps and stairs   117
6.14 Roofs   118

7 Structural design   119
7.1   General   119
7.1.1 The partial factor method   119
7.1.2 Bridges   119
7.2   Preliminary design sequence   120
7.3   Principles of stabilisation   121
7.3.1 Bracing forces   122
7.3.2 Means of providing stability   123
7.3.3 Triangulation   124
7.3.4 Portal principles   124
7.3.5 Diaphragms   125
7.4   Analysing the main structural systems   125
7.4.1 Preliminary design   125
7.4.2 Detail design   126

Contents

7.4.3 Form-specific analysis aspects   126
7.4.4 Preliminary calculations   127
7.4.5 Stability of arches   127
7.5   Continuing the analysis: validation stage   127
7.5.1 FEM calculations   128
7.6   Fulfilling serviceability conditions   130
7.6.1 Limiting values for deflections   130
7.6.2 Limiting vibrations caused by pedestrians   131
7.6.3 Large pedestrian bridges, and vibrations experienced
       by pedestrians on vehicular bridges   131
7.7   Connection design   132
7.7.1 Connection design fundamentals   132
7.7.2 True pins   133
7.8   Fastener and connector design   134
7.8.1 Choice of fasteners and connectors   134
7.8.2 Corrosion protection   135
7.9   Validating connection design   135
7.10   Durability of connections   137
7.10.1 Moisture movement of timber around multi-dowelled plates 138
7.11   The final details   139

8 Conservation, maintenance and repair   140
8.1   Importance of inspection and maintenance   140
8.2   The influence of the structural form   140
8.3   Maintenance and construction specifications   141
8.4   Erection sequence: inspection and maintenance
       database   142
8.5   Construction stage   142
8.6   Inspection and preventative maintenance   142
8.6.1 Techniques   143
8.6.2 Recording   143
8.7   Inspection   144
8.7.1 Pre-inspection evaluation   144
8.7.2 Field inspection   144
8.7.3 Substructure inspections   145
8.7.4 Superstructure inspection   145
8.8   Vibration testing   146
8.9   Reports and records   146
8.10   Repairs   147
8.10.1 Wearing surfaces   148
8.10.2 Surface treatments   148
8.11   Mechanical repairs   148
8.11.1 Member augmentation   148
8.11.2 Metal plates and reinforcements   148
8.11.3 Stress laminating   149
8.12   Adhesives for strengthening and stiffening   149
8.13   Bridge strengthening example: Tourand Creek Bridge,
       Winnipeg, Manitoba, Canada   149
8.13.1 Appraisal   149
8.13.2 Testing   150
8.13.3 Preparation   151
8.13.4 Execution   151
8.13.5 Advantages and savings   152

9 General case studies and final recommendations   153
9.1   Interpreting timber bridges   153
9.2   Pont de Crest: a contemporary environmental
       approach in central France   155
9.2.1 Conclusions   156
9.3   Punt la Resgia: a modern covered arch design in
       Switzerland   158

9.4   Toijala, Tampere: a bridge with distinctive trusses for cyclists and pedestrians – Finland   160

9.5   Ollas, Finland: a pedestrian and cyclist flyover using classic glulam arches   162

9.6   Flisa Bridge: a record-breaking trussed road bridge in Norway   163

9.7   Almorere pylon bridge: a pedestrian and cycle crossing in the Netherlands, using innovative timber engineering   165

9.8   Passerelle d'Ajoux   167

9.9   Tharandt Forest Botanical Garden: tree-top walkway   169

Notes   171

Foreword

Timber construction is undergoing a significant revival thanks to the combination of several key factors. Timber really does grow on trees! Not only is it renewable, offering a potentially unlimited source of supply, but it also grows using energy from the sun, making it far less sensitive to the price fluctuations of electricity. Increasing automation in harvesting, sawmilling and fabrication is also helping to reduce the costs of production. At the same time, designers have been re-learning the skills of designing and detailing in this natural cellular material. Timber is of particular value in bridges, where its high strength-to-weight ratio enables the prefabrication of large elements, helping to reduce the costs of erection, which form so significant a part of any bridge design.

Given all this, it will come as no surprise that recent years have seen a major renaissance in timber bridge engineering. *Timber Bridges* could not have been written at a better time. The book covers the entire range of issues that designers will need to address, and it is also generously illustrated, showing what has already been achieved to date.

It would obviously be wrong to suggest that timber bridges do not present their challenges, particularly to ensure an adequate design life. While hundreds of timber bridges do survive from the 19th century and earlier, these generally rely on roofs to keep the main structure dry in order to prevent fungal attack. Subject to maintenance of the roof, the main structure has an almost indefinite life. The 20th century saw this costly physical protection replaced by the cheaper chemical protection offered by preservatives. However, preservatives are by their nature toxic, and even the most toxic preservatives (today, increasingly restricted by legislation) cannot achieve much more than a 30–50 year life for timber elements that are fully exposed to the rain.

Recent years have therefore seen a re-evaluation of the benefits of physical protection, combined with careful detailing to avoid water traps in order to keep the timber as dry as possible. A 'belt and braces' approach is the most effective way of dealing with waterproofing. This accepts that the first line of protection might fail and provides an alternative line of attack, such as the use of a preservative, a naturally durable timber or simply ventilation and drainage below the upper layer of protection, to ensure that any water which does leak through can quickly drain away and evaporate.

Methods of protection against wetting, therefore, form one of the most important parts of this book. The manner of physical protection will have an obvious impact on the appearance of the bridge, and *Timber Bridges* emphasises that the choice of structural form and the method of protection and maintenance need to be considered from the very start of the design. The book also covers the choice of species, treatment techniques, connection detailing, design of decks and parapets and maintenance issues.

As the many varied examples in the book show, the simple options of beam, truss, arch and catenary belie the tremendous variety of aesthetic forms that can be achieved in timber. For the first time, the book brings together all the information the designer needs to develop a robust timber bridge design. By explaining how the various design criteria can be addressed and by showing what has already been achieved across the world, it is hoped that this book will inspire many more designers to venture into the field of timber bridges.

Andrew Lawrence, MA (Cantab), PGCDMM, CEng, MICE, MIStructE.
Associate at Arup Technology and Research

1 Benefits of timber bridges

1.1 Essentials

Bridges in timber for pedestrians, cyclists and equestrians are enjoying a significant revival. Recently, they have also come to be regarded as suitable for longer spans and heavier loadings, including regular vehicular traffic (*Figure 1.1*). The high strength-to-weight ratio has long been recognised, and for this reason loads upon new or existing foundations are kept to a minimum, a motive for several of the projects illustrated herein. With timber engineering, off-site fabrication has always been completely normal, ensuring considerable versatility. Prefabrication in sections takes place in factory environments with strict levels of quality control fully addressed by European Standards and codes. Large components or even entire bridges are often rapidly lifted into place and assembled, as in *Figure 1.2*, and again as exemplified in later chapters. Ease of erection is especially significant in situations where minimal disruption of traffic is essential.

Timber is ideal for applications where aesthetics are important, and the awakened public and professional awareness of its sustainability has undoubtedly increased for these reasons. There are suitable forms of bridge for both rural and urban environments, and a range of surfaces, colours and types of finish are provided.

Figure 1.1 Timber provides excellent bridges for pedestrians, cyclists and equestrians. It is also a suitable material for longer spans and heavier loadings, including regular vehicular traffic.
A: Crossing the river Neckar, Remseck, Germany.
B: Vihantasalmi Bridge, Mäntyharju, Finland
Photo A: © STEP
Photo B: © CJM

Figure 1.2 By nature lightweight and suitable for prefabrication, timber bridges for footpaths and cycleways minimise time on site. Here a construction in 2008 at Traunreut, in Germany, has round-turned glulam and protected rectangular-section glulam stringers with round hollow section steel. A roof fabricated from innovative structural laminated panel material completes the project
Photos © F. Miebach/Schaffitzel

This publication contains comprehensive information on timber bridges of moderate span, suitable for relatively light types of traffic, since these represent the majority of present applications in the United Kingdom. However, it also exhibits large modern engineering achievements beyond these shores. In interviews with construction professionals, and during presentations at training events, astonishment is often expressed at the scope and scale of timber bridge developments elsewhere. The advent of the unified suite of structural Eurocodes, including one specifically addressing timber bridges, has removed a major obstacle in the eyes of British engineers and potential approvers. There is a perception here that timber bridges lack any prospect of longevity, yet the evidence of still-standing medieval structures disproves this notion. Within this book there are examples of bridge accomplishments that embrace the very latest timber technology and engineering principles, and that are confidently expected to fulfil the demands of formally defined design lives of up to 120 years.

In the last 25 years, there has been considerable international co-operation on modern timber bridges amongst experts and specialist fabricators. The Nordic Timber Bridge Project and other similar initiatives have resulted in hundreds of new bridges in the northern latitudes, where forest products are naturally abundant. While the majority of these are pedestrian types, new heavy-traffic bridges have also been installed. The advent of efficient supply chains linked with electronic access to technical information has led to opportunities for timely deliveries to the United Kingdom of prefabricated components similar to those used in Denmark, Finland, Norway and Sweden. This can involve energy-efficient methods of supply, including rail and water transportation of large, sectionalised parts for bridges. The Nordic project has included scientific life cycle assessments that support environmental declarations, so these are not just vague 'wood is good' statements. Elsewhere in Europe, timber engineering activity has also grown, and a number of illustrations shown in this publication relate to bridge projects in France, as well as in the German-speaking countries of Europe. In North America and in the Antipodes, renewed timber bridge programmes are also well established, with published design documents available, to which roads and traffic authorities routinely make reference.

United in their conviction as to the efficiency, durability and other benefits of these bridges, timber engineers around the world nevertheless need to respond to differing national and regional priorities and motives. To those having some familiarity with the subject, these may be recognised in many of the illustrations used throughout, but to illustrate this variety further, selected case studies are presented in Chapter 9.

1.2 Durability

By far the most effective way to ensure durability and ease of maintenance of any timber structure is to incorporate protective design measures at the initial design stage. In Chapter 2, examples are given of timber bridges that have lasted for centuries, due to protective design features – in particular, covers over part of the structure, or indeed a complete 'overcoat' in the form of a roof. Designing for durability is examined in detail in Chapter 3, while materials choices and their factors are discussed in Chapter 4. At this stage, it is emphasised that the species of timber selected for bridging should always possess a degree of natural durability, even when the options of including preservative treatment or adding full or partial coverings are chosen.

At an early stage in the concept design, it is essential to adopt a decisive protection strategy. The three-pronged approach entails careful timber selection and specification; surface treatment and possibly in-depth treatment with preservatives; and physical protection of key parts of the bridge using measures detailed herein. Standard durability classifications are provided for all of the commonly available bridge timbers. These ratings are also linked to hazard definition classifications, which are discussed in Chapter 3. The primary choice of timber species and material format is crucial. It interrelates not only with technical design matters such as structural calculations, for which it is necessary to select a strength class or a nominated timber species, but also with issues such as available cross-sectional sizes, lengths, and amenability to pressure preservative treatment if stipulated.

1.3 Benefits

Using timber in construction in general, including bridges, has important environmental benefits, including:

- very low embodied energy during manufacture;
- low mass, lessening transport and foundation construction energy;
- positive carbon balance;
- the opportunity to use a perpetually renewable material – hence encouraging the sustainable management of forests and woodlands.

The Pont de Crest (*Figure 1.3*) is a recent example of a bridge for which timber was specifically chosen on environmental grounds. Because of faith in the quality of the local forest products and the skills of their carpenters, timber was the preference of the citizens of Crest in Drôme, France for this attractive and successful structure.

Figure 1.3 The Pont de Crest constructed in 2001 to cross the river Drôme, in France. With a length of 94 m, this bridge is distinguished by its tree-like architecture, conceived to serve the community whose livelihood is heavily dependent upon forestry and its products. The beams, struts and a composite under-deck diaphragm are all produced from local Douglas fir. Several innovative features are included in the connections and in the protective design details. The bridge carries a local secondary road and is designed for a 10-tonne capacity, although normally vehicles of only 3.5 tonnes are permitted
Photo © CJM

1.3.1 Sustainability

Uniquely amongst structural materials, timber is completely renewable by nature, bringing clear environmental benefits. These include a very low embodied energy content in converted products, especially if they are derived from relatively local sources. Applications such as bridging provide skilled design and manufacturing opportunities for rural and regional communities, leading to the expansion of forest and woodland protection and viable cyclical harvesting. While they are still growing vigorously, trees absorb carbon dioxide, but this benign effect is only experienced if the natural resource is correctly protected, utilised and renewed when the time is right.

Well-managed timber production has a tradition spanning back for centuries in Britain, and in most other parts of Europe, as well as the whole of Scandinavia, and this can be sustained through innovative and efficient practices. For several centuries, softwoods have provided the 'traditional' structural timber in most parts of Europe, but over the past few decades there has been a strong revival of interest in the use of broad-leaved timbers such as oak, ash, beech and sweet chestnut. Specifiers can have a major influence on such choices. Careful design and detailing reduces reliance upon chemical preservatives. The choice of durable timbers for exposed applications such as timber bridges can also keep treatments and finishes to a minimum.

Recently, all of the Forestry Commission woodlands in England, Wales and Scotland were assessed against FSC- and ISO-recognised schemes. This showed that the best possible practices of management and replanting have been taking place. This excellent record is reflected also in the private sector of arboriculture in Britain. In the Nordic and North American regions, FSC and PEFC schemes apply comprehensively, and their abundant supplies of forest proffer considerably greater opportunities, both in timber bridging and also in building design, than are currently realised.

1.3.2 Forestry benefits

Engaging sustainable forestry involves making conscious decisions on the future of the land and biosphere, paying careful attention to four fundamental aspects:

1 Physical, biological and ecological considerations.
2 Social, political and cultural aspirations.
3 Economic and financial opportunities and constraints.
4 Growing, managing and fulfilling technological aspects – at all stages in the forestry–timber chain.

With very few exceptions (*Figure 1.4*), the most durable species of timber are obtained from tropical sources. Those readily available in Europe for bridge construction are described in Chapter 4. Reliance on the natural durability of the heartwood is only one of the three main avenues leading towards the lifetime performance now expected of permanent bridges. Nevertheless, the larger cross-sections and greater available lengths of tropical timbers mean that these types are particularly suited to civil engineering applications. There may still be some confusion about the correctness of felling trees for timber generally – however, this activity is not only essential, but also complementary to the natural cycle of regeneration, growth and maturity. Sustained management rather than exploitation is the proper aim, and this needs to apply especially in tropical regions where, in the recent past, this has not always been the case.

Figure 1.4 Taking advantage of sustainable forestry – in this case by using a durable broad-leaved temperate timber available locally (sweet chestnut heartwood) – for a 20 m span arched footbridge, the Passerelle d'Ajoux, in the Ardèche, France
Photo © J. Anglade

In parts of the world such as the lower Amazon basin, deforestation remains a serious concern. Elsewhere, there may also be disadvantages with the excessive planting of economic trees, such as rubber and oil palms, when these form vast monocultures – even though they are a means of sequestering carbon, as well as providing the crops concerned.

On a positive note, in developing countries, where most of the world's hardwoods grow, forestry and forest industries continue to offer long-term opportunities for local populations. As well as maintaining existing forests and woodlands, possibilities include afforestation of new lands and the introduction of various forms of social forestry such as agroforestry (*Figure 1.5*). These activities may be accompanied by the introduction of appropriate technologies in the wood industries, whose employees may thus remain settled rather than drifting into cities. Such products may have significant added-value in the country of origin, and specifying them to become chain-of-custody certified hardwood members and components in structures is greatly preferable to boycotting these sources. In view of the excellent quality and durability of selected tropical timbers, this attitude is far from risking the destruction and impoverishment of the land and its occupants.

Figure 1.5 An example of sustainable tropical forestry – iroko, planted in 1958 and photographed approximately 32 years later. This is part of an agroforestry development in Cameroon, West Africa, illustrating how success requires patience, and that timber-bearing trees can be raised alongside food crops, both to the benefit of local populations
Photo © CJM

Figure 1.6 A large timber road bridge at Evenstad in Norway, taking the traditional form of bowstring trusses; crossing the river Glomma in Hedmark County, where all of the forestry operations are performed to the highest environmental standards; through several stages, locally harvested European redwood logs are converted into durable double-treated glulam timber components. This structure has five spans each of 36 m
Photo © CJM

Figure 1.7 The Passerelle Pinot, a pedestrian crossing bridge at Blagnac, Haute-Garonne, Midi-Pyrénées, is an example of a mixed timber and steel bridge in which the protection of the 36 m span glulam tied-arches was taken into account in the architectural concept right from the start, resulting in a popularly acclaimed work of engineering art
Photos © Alain Mochi via Structurae

1.3.3 Appearance

All of the bridges illustrated in this publication have been selected to illustrate the beauty of timber structures (*Figure 1.6*). In approaching a new design, the aesthetics need to be considered both from a distance and at close range. The overall impact of the bridge needs to be conceived in relation to its environs, whether they are in the natural landscape or in a town or city. Use of the bridge needs to be a pleasing experience for the public and for its visitors, and attention should be given to the textures and colours of the materials used. The protection of the bridge, based on a 'belt and braces' principle – i.e. always including a fall-back protection route in case the primary one begins to fail or require attention – should be included within the architectural concept right from the start (*Figure 1.7*).

Consideration also needs to be given not just to how the bridge will look when completely new, but also to how it will appear after a reasonable passage of time, given the degree of maintenance that is likely to be achieved. Since this may vary, according to the type of bridge owner and situation, factors such as amenability to weathering, with or without applied finishes, need to be assessed.

Stunning bridge forms, relating to the vocabularies of contemporary architecture, are feasible. To replace an older timber bridge, or indeed a more traditional type in any other material, local wishes may indicate a traditional architecture and form for the new structure. Where a bridge is to complement a new landscape, then one of the exciting contemporary forms shown throughout this publication may be appropriate.

1.3.4 Low mass

Historically, the developments that took place in long-spanning masonry and reinforced concrete structures were made possible through skilful and impressive timber falseworks, themselves spanning considerable distances and carrying heavy loads. Knowledge and experience of falsework design fed back into timber bridging, where the high strength and low mass of wood gave obvious benefits. Nowadays, structural timber composites are

available for bridge engineers, where the manufacturing processes eliminate most of the natural defects, substantially reducing the statistical variability and increasing stiffness through homogeneous drying.

Timber often provides the opportunity to re-use existing foundations and piers for a more modern and durable structure, sometimes positioned with greater clearance above the route or watercourse to be crossed. Projects where this has been an important consideration include the Flisa Bridge in Norway, shown in Chapter 9.

Several of the examples included herein describe transportation and erection techniques, since timber engineering is essentially a factory or workshop prefabricated technology, meaning such matters need to be considered at a very early stage in the project. The use of glued laminated timber (glulam), laminated veneer lumber (LVL) and other forms of structural timber composites (STCs) is described in Chapter 4. The same manufacturers who produce the STCs have in-house and closely associated specialists who can provide independently assured services for detailing, delivery and erection. For these vital project stages, techniques are essentially similar to those for steel structures.

Attention needs to be paid at the concept design stage to the choice of site erection nodes and the division of major components into vehicle loads or packs for delivery. However, the lighter mass of timber gives greater scope for decisions over cranes, temporary towers and other forms of lifting, while the option of division of prefabricated and trial-assembled elements permits delivery to the most remote sites.

1.3.5 Low whole-life costs

Timber bridges bring further user benefits in having low whole-life costs. Careful design and detailing, surface treatments and the use of durable or preservative-treated timbers will ensure a long service life. Regular maintenance will keep down costs, and bridges can be designed in such a manner that this is a simple process. Attention to both concept and detailed design at the early stages prevents the costs of major replacements later on. Less major parts may intentionally be designed to be easily replaceable after a shorter design life than the principal structure.

1.4 Applications

As we have already noted, timber is ideal for applications where aesthetics are important. Moderate spans, for relatively light traffic, characterise most of the present applications in the United Kingdom. However, engineering achievements elsewhere should stimulate the imagination of designers and owners towards greater ambition. Examples collected in the case studies in Chapter 9 are preceded by a brief comparative analysis of international trends. Throughout this publication other illustrations will also be found to fit this scheme.

To focus on the preliminary steps necessary to develop a design proposal, a selection of common applications is taken forward in Chapter 5. The non-exclusive list used there is as follows:

• Footpaths, cycle tracks and bridle paths over roads – often providing statutory rights of way.

- Railway crossings – usually just for pedestrians and cyclists, sometimes including equestrians and light vehicles.
- Footpaths, cycle tracks and access routes across water features, for example in areas of outstanding natural beauty, national parks, nature reserves and sites of special scientific interest.
- Recreational and leisure activities – very varied applications often favouring timber – includes stately homes and gardens, garden festivals, parks, botanical gardens, golf courses and seaside attractions.
- Carrying roads over other highways, across land or water features or railways.

Before proceeding towards concept design, we examine in Chapter 2 the range of possible structural forms, the vital approaches to ensure durability – Chapter 3 – and the main materials options within the generic timber and wood-based field (Chapter 4).

1.5 Modern designs

In modern timber bridge design, there is a wide choice of structural form. Primary beams, arches and trusses are all quite common. Cable-stayed and suspension bridges are built, with structural timber playing a major role.

Tension-ribbon and bascule types are also used (*Figure 1.8*), whilst stressed laminated decks, discussed in Chapters 6 and 7, further increase the engineer's repertoire. The selection of the form and the types of component will depend on many factors, including the nature and profile of the site, the loadings and spans, as well as the required clearances.

During the past four or five decades, connection design has undergone a period of great change and improvement. Extremely efficient, concealed and durable structural nodes are now being designed using procedures contained in BS EN 1995-1-1 and 1995-2 (timber structures – general and timber structures – bridges, respectively) and this topic comprises a major section of Chapter 7.

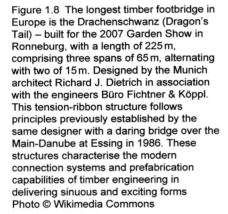

Figure 1.8 The longest timber footbridge in Europe is the Drachenschwanz (Dragon's Tail) – built for the 2007 Garden Show in Ronneburg, with a length of 225 m, comprising three spans of 65 m, alternating with two of 15 m. Designed by the Munich architect Richard J. Dietrich in association with the engineers Büro Fichtner & Köppl. This tension-ribbon structure follows principles previously established by the same designer with a daring bridge over the Main-Danube at Essing in 1986. These structures characterise the modern connection systems and prefabrication capabilities of timber engineering in delivering sinuous and exciting forms
Photo © Wikimedia Commons

1.6 Architecture

The desired architectural statement is a fundamental aspect of form selection. As indicated above, the protection scheme is also integrated right from the start of the project, rather than being included as an afterthought. But the designer seldom has a totally free rein on the general arrangement, since the choice depends upon practical factors such as the general site plan, terrain profile and route alignments. A successful bridge design should resolve both the aesthetic and the practical requirements in a manner that is subliminally fulfilling to the observers. With regards to further architectural choices, a bridge may be sought which provides a landmark structure, or the brief may be to blend into the surroundings inconspicuously, or in a manner in keeping with surrounding buildings. Prompts towards the final scheme design may also be derived from the historic, cultural or contemporary nature of the district in which the bridge is to be located.

1.7 Generic timber information

One advantage timber has over almost all other materials is that trees are a living, and therefore renewable, resource. Timber and other forest products can be grown and harvested in just the same way as any other form of agriculture, though on a longer time scale.

The commercial division of timbers into hardwoods and softwoods has evolved from traditions when the timber trade was dealing with a limited range of species. Today, however, this division bears no relation to the softness or hardness of the timber. Softwoods are produced from coniferous or cone-bearing trees that have needle-like leaves and are mostly evergreen (e.g. pines and yew). Hardwoods are produced from broad-leaved trees which produce seeds contained in an enclosed case or ovary (e.g. an acorn or walnut).

As a natural material, the efficient utilisation of timber requires some form of selection and grading. An understanding of the characteristics and properties of timber as a raw material will enable the designer or user to ensure that timber is used to best effect. Chapter 4 of this book provides useful data on the properties of many timbers suitable for bridges, while the online technical information library of the Timber Research and Development Association includes a wide range of information sheets for specifiers and designers on the effective use of timber.

1.8 New materials

In extending the scope of materials available to the modern timber engineer in terms of length and cross-section, glulam and other STCs offer important opportunities. These materials are manufactured in quite specific formats and from very particular timber species, so it is necessary to choose them carefully, since certain types have either increased natural durability or are capable of receiving preservative treatment. These subjects are examined in Chapter 4. Throughout this publication many of the illustrated examples and case studies feature innovative materials and techniques.

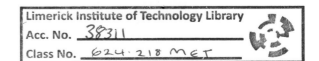

2 The evolutionary development of the timber bridge

2.1 General

First, some general classifications of bridges are required, but beginning purely with supposed chronological development is unsound. An alternative is to classify through the type of structural arrangement, and although this also needs to be done carefully, taken in conjunction with broadly indicative evolutionary examples, it provides a better framework.

There can be said to be five essential bridge forms and types:

1 beams
2 cantilevers
3 suspended structures
4 arches
5 trusses.

As shown in *Figure 2.1*, each of these can be related to various forms occurring in nature. It is evident that the truss is the most sophisticated type, requiring ingenuity to ensure adequate, reliable connections. To find it in nature we generally have to look inside animals' bodies, where there are ligaments as well as bones – giving the essential members and tied joints of the truss. Not surprisingly therefore, this relates to our understanding of the broad chronological status of human bridge engineering.

Figure 2.1
Photos © Wikimedia Commons

1 Beams	2 Cantilevers	3 Suspended structures
Fallen tree spanning a stream; a beaver lodge. 	Natural rocks with balanced boulders; horizontal tree branches. 	Spiders' webs; lianas and vines 

4 Arches	5 Trusses	
Weathered arches in natural rock formations. 	Growth structures, e.g. skeletons (plus ligaments) and other internal natural forms in plants and animals.	

Beams, on the other hand, are broadly observed in nature – for instance, in a fallen tree already conveniently spanning a stream. It would only have required a modicum of initiative to trim and rearrange one that was in about the right place, or even to fell a new tree using stone or bronze tools. Beaver lodges are large and complex items of animal engineering and, bearing in mind the wider past habitat range of these creatures, they must often have provided humans with useful ready-made beam bridges.

Cantilevers can be seen amongst natural rock formations where there are balanced boulders. Horizontal tree branches also have the necessary anchorage already in place, taking the weight of the climber back to the trunk. Suspended structures are easily noticed amongst spiders' webs and – on a larger scale – in lianas and vines. These, like the natural beams, would sometimes need hardly any rearrangement. With a little more ingenuity, they can easily be attached to rough posts driven into the banks of a river, the anchorages further loaded if necessary with piles of stones. In many remote regions, plaited natural materials of high tensile strength, such as bamboo and rattan, are still serving to provide crossings in a similar manner. Arch formations providing natural rock bridges are well known in diverse parts of the world, and some have become famous tourist attractions.

The principles of structural mechanics can be added to the taxonomy, but this needs to be done cautiously, because amongst the five forms, there are numerous hybrids and modifications. For example, there is hardly an arch that does not, under some circumstances, experience flexural actions, while even slightly bowed beams exert a modicum of horizontal thrust. A tension-ribbon – the purest bridge in terms of action – will need to be designed against lateral effects due to wind – as realised by spiders!

So, taking into account the extreme simplifications that such reservations necessitate, one can introduce the three principal strain types of structural mechanics, namely:

- tension (T)
- compression (C)
- shear (S).

These may then be compiled in a classification of human-built structural bridge forms, as in *Table 2.1*, from which sub-types that seldom occur in timber are omitted.

Why is classification by ancestry dubious? At an early stage in evolution, mankind must have crossed naturally occurring bridges. But by definition 'pre-history' is unknown, so who can tell which type of bridge was first constructed artificially? Certainly, this entailed stonework and timber, probably in the form of the log beam bridge! Also, perhaps, other natural plant materials such as lianas and vines were used at a very early stage in the activities of *homo sapiens.*

2.2 Beams

Bronze-aged causeways such as that from Peterborough to Whittlesey Island, excavated by Pryor, exemplify significant timber planked structures, with transoms supported on driven piles and connected by simple mortises.

Table 2.1 A bridge classification by structural form

Primary form	Dominant action effects	Sub-types and extensions of principle
Beams	Flexure i.e. $T + C + S$	Girders with parallel chords[†] Plates/slabs Under-tied beams Multi-span trestles – with or without fan struts Under-strutted beams and portal frames
Cantilevers	F at cantilevered span; S at encastré locations; • essential to have massive abutments	A cantilevered stack of solid, built-up beams Latticed cantilever bridges are unusual in timber engineering, but they are indeed feasible 'Suspended' central spans of continuous sets of beams are quite common, and here of course the side sections provide a cantilevered extension to support the centre
Suspended structures	T with essential anchorage at supports	Simple suspension bridges over gorges – natural abutments Tension-ribbon, usually on high banks and/or with braced towers/trestles The classic suspension bridge, supported from masts/towers Cable-stayed bridges
Arches	C with essential abutments to resist thrust	Tied-arch – alternative to abutments Latticed arch Simplified arch, e.g. A-frames/kingposts[‡] Decks of arched bridges may be positioned at various levels within the structure – see Chapter 5
Trusses	Direct actions within essentially quite slender individual elements, $T + C + S$ according to the truss analysis[§]	There is a great diversity of truss types – trussed bridges in timber being common and evolutionary. Some are illustrated in this chapter, more subsequently. Bowstring trusses are closely similar to tied-arches, while some types of historic timber bridge are effectively arch-stiffened frames or trusses

Notes

† Parallel-chorded girders could be treated as a type of truss. However, some types of beam analysis apply the 'truss analogy' – upper fibres, *compression;* lower, *tension;* linkages, *shear.*

‡ Clearly kingposts might be regarded as trusses, demonstrating the ambiguity discussed above.

§ Consider the Vierendeel truss (or is it a girder?) – internal triangulation is normally considered a primary feature of the truss, but the Vierendeel has none. Vierendeel forms are used in timber.

Figure 2.2
Caesar's Bridge on the Rhine – lecture drawing, Sir John Soane
Photo © Sir John Soane Museum

In works still accessible today, there are clear records of Roman timber beam bridges. For example, Julius Caesar had a military bridge built across the Rhine in 55 BC (*Figure 2.2*). Based on the primary texts, Alberti (1404–1472) and Palladio (1518–1580) describe this structure in detail. Caesar also writes about a large example in Italy, while other classical works discuss the Pons Sublicans, an important entry into the City of Rome. This was underpinned by timber grillages (a possible explanation of its name) and probably had a significant superstructure of timber at the early stages of its multiply renewed life. There is archaeological evidence that the Roman bridge in London was by no means a crude or simple structure, while there are also traces of evidence for good quality Romano-British timber bridges at Newcastle, Chester and on the Severn and Wye. On the continent of Europe, piles from Roman times are preserved from the large and impressive bridge over the Mosel, at Trier, for example.

On the Thames in London, late Saxon board-walks and jetties have been revealed in excellent condition and recorded by Milne. The oldest timber bridges that are still in use in Europe date from around the 14th to the early 16th centuries. Many of these are covered beam bridges, formed with multiple spans, using intermediate supports comprising trestles (*Figure 2.3A*). They owe their longevity to the simple protective device of the roof. A fine example of this ancient type is in Lucerne, where the Kapellbrücke can still be appreciated, thanks to its most recent restoration in 1994. Although the superstructure had been damaged by fire, original timbers were rescued, and many of the oak piles sunk into the lake bed to support the trestles are also original. Andrea Palladio, the great architect of the renaissance, whom we shall discuss presently in connection with truss bridges, also left us the simpler structure at Bassano (*Figure 2.3B*); re-built several times after wartime destruction, but still true to the original design.

The unification of Switzerland as a nation, and the integration of its federated states, gave it commercial impetus and increased wealth. Bridges that are very important both historically and technically, are those built by the State of Berne during the 16th century. These include structures at Neubrugg (1532), Gummenen (1555), Wangen (1559) and Aarberg (1568). Most are still in good condition, having their original main elements, and some still carry vehicular traffic. Actually, some of the multiple spans of these Bernese bridges are formed as hybrid trussed-arched frames. This is a transitional historic type that for convenience we will examine later.

Figure 2.3 A: the Kapellbrückein Lucerne, dating from 1333, and recently restored. B: in Bassano, this trestled and roofed beam bridge, built in 1558–1560 and repaired several times following wars (most recently in 1947), stands testament to Andrea Palladio, the great architect of the renaissance
Photos © Wikimedia Commons

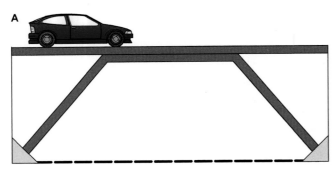

A

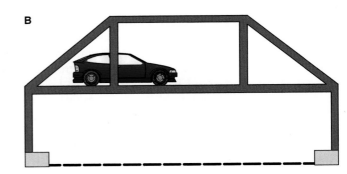

B

2.2.1 Under-strutted beams and frames

The use of under-strutting to extend the span and load-carrying performance of simple beams is a centuries-old technique that still sometimes has virtue. Generally, it is only possible where there is a suitable clearance under the deck. If used for a bridge under which powerful water flows, there is a danger of the structure being damaged or destroyed by flooding.

2.2.2 Railway trestles

At the start of the Industrial Revolution, timber was of considerable importance as a bridge-building material in Britain, as was also the case elsewhere in Europe.

Figure 2.4 Traditional timber bridges dating from the 16th century are classified in German language publications, as 'Sprengwerk' (A), or 'Hängewerk' (B). The 'rod polygon' hängewerk (B), in this case a classic 'queenpost truss', can be thought of as the precursor to fully-formed arches. See, for example, Lucerne Spreuerbrücke, (Figure 2.5), which contains both types
Drawing © CJM/TRADA Technology

Figure 2.5 The Spreuerbrücke, Lucerne, illustrates the transition from 'hängewerk' polygonal frames to curved, bent-laminated arches.
A: entry to the bridge;
B: painting – 'Totentanz' by Kaspar Meglinger, 1626;
C: the bridge chapel, once a common feature;
D: Hängewerk frame;
E: arch of six bent-pine laminations each six inches thick;
F: the covered bridge at Bad Säckingen, the longest of its type in Europe, shows a series of developments – founded in 1272, the current structure is a re-build (following wartime destruction) of a truss type similar to a Howe, but with all-timber members; its length is 202.9 m, and it is now a cycle- and footbridge
Photos A–E: © CJM
Photo F: © German Wiki

Figure 2.6 A: an all-timber Brunel viaduct at Penponds Viaduct nr. Cambourne, 1884. Many taller structures were also built on stone masonry piers and these are famous, but this example has timber piles as well as superstructure.
B: Barmouth Viaduct, 1921
Photo A: © The Cornish Centre Collection
Photo B: © Project Guttenberg

Isambard Kingdom Brunel was the most gifted of the great engineers engaged in the rapid 19th-century expansion of British, and subsequently international, railways. He placed fitness for purpose, economy and rapid availability above all else. Hence, he used treated demountable timber elements to build major viaducts for fast railway development (*Figure 2.6A*). On the South Devon, West Cornwall and Cornwall railways routes, 64 large timber viaducts were built between 1849 and 1864. Simple, multi-span post-and-beam structures used two-layer laminated timber beams. Timber truss bridges of various forms were constructed, and perhaps most famously, 'fan' forms of propped multiple-span beams created the viaduct trestles. In South Wales, several of these continued to be in service for up to 100 years. Brunel often had these designs checked by full-scale prototype testing involving large segments of the structure.

It is less well known that Brunel also had a number of able timber engineering peers. Details, contemporary illustrations and primary reference leads are given by Booth. Examples of the engineers and their timber forms are Cubitt (propped beams, extended queenpost trusses); Gooch (double polygonal arches); Green (multiple span, elevated laminated arches); Stephenson (polygonal arches); and Valentine (laminated bowstring trusses).

2.2.3 North American trestles

When these were first built, huge volumes of inexpensive timber were employed to bring the railways into operation as rapidly as possible. As already noted, Brunel and his contemporaries had already adopted a similar principle in Britain. A small number of early North American examples remain in service, carrying locomotives (*Figure 2.7*). More have been converted for use as recreational routes, providing access for walkers, joggers and cyclists.

Figure 2.7 Wilburton Trestle, Washington State, one of the many North American timber rail trestles; this structure is still carrying heavy rail traffic to and from the Boeing aircraft factories
Photo © Brianhe

2.2.4 Conserving trestles

Trestles form conspicuous features in the landscape, and local communities have often championed them through an increased awareness of the merits of conservation, and a local pride that can become very fierce. At Wickham Bishops, near Maldon in Essex, a timber railway trestle has become a 'Scheduled Ancient Monument', and is an interesting feature of one of the popular local access routes for walkers and cross-country cyclists. This structure, built in 1847, was part of the former Witham to Maldon Branch line. It comprises heavy square-section Baltic pine timbers in characteristic braced post and longitudinal beam formation. Six such viaducts were originally included along the route. The measured and documented survey is archived by Essex County Council, while some illustrations of the sensitive repair work are given in the above referenced work by DeLony.

In the north east of Scotland, 24 km south of Inverness, there is an extremely rare instance of a British timber trestle that is still in use for main-line rail traffic. According to the Scottish Grade-A listing, the structure is known as Aultnaslanach Viaduct, Moy, but because of its award-winning conservation project, completed in 2002, it is generally known as Moy Viaduct. Probably one of the reasons for the survival of its five spans since 1887 is the good behaviour of its timber piles over all those decades sunk in an extremely acidic peat bog. This structure has now been discretely strengthened with a concealed steel frame to support the weight of modern trains. In situ timber treatment and partial replacements using appropriate species led to a British Construction Award citation in 2004 for 'a careful, sensitive treatment of a valued listed building'.

Finally, a trestle bridge example from Victoria, Australia (*Figure 2.8*) illustrates how local passions can become inflamed over bridge conservation issues. In the late 19th century, Barwon Heads was just a small fishing community, but during the 1920s and 1930s it became a popular holiday resort within reach of Melbourne. The estuary bridge was opened in 1927, crossing the Barwon river where a ferry service had previously existed. It has formed the backdrop for several films and television documentaries. In 2005 the government of Victoria expressed

Figure 2.8 Barwon Heads Bridge – a celebrated conservation cause in Victoria, Australia
Photo © Wikimedia Commons

concern over its load-carrying capacity, and installed a 15-tonne limit, pending consultations. Originally, the intention was simply to dismantle and replace it, but in 2006 this was successfully opposed by a 'Friends of Barwon Heads Bridge' non-profit pressure group. The reconstruction project became a celebrated cause, and the Victoria roads authority acknowledged that 'the historic Barwon Heads Bridge will … retain its iconic status well into the future'. Later, in a compromise that is still not altogether accepted by all concerned, the plan is for the road bridge to be rebuilt using a mix of the original hardwood timbers and replacements to match, while a new footbridge for pedestrians, cyclists and anglers will be constructed nearby.

2.2.5 Contemporary beams
Using modern timber engineering materials and the protective design techniques described in Chapter 3, conventional beam bridges of up to 42 m span are possible (*Figure 2.9A*). Where the site and structural conditions permit, under-tied beams are another economical type (*Figure 2.9B*), which illustrates a contemporary design to complete a pack-horse lane, which also contains several historic stone hump-backed arches.

Block laminating is a specialist manufacturing technique that has recently been introduced in Germany to bond together a series of normal glued laminations into massive sections. These may be formed into complex shapes and curved plan-forms, offering opportunities with shallow curved decks and special profiles. This technique is discussed further in Chapter 4. A relatively simple beamed footbridge of this type is illustrated in *Figure 2.10*.

Figure 2.9 A: the 'Mursteg', a covered beam bridge with integrated, sheltered approaches. Crossing the river Mur at Murau, Austria, the main span is 22 m. Each of the lower chords of the glulam beams are pre-tensioned with an embedded steel tendon, and the parapets fulfil a lateral stabilising function as well as offering protection. The deck is stressed laminated.
B: Pont Bouix, France – the under-tied hump-backed form was conceived specifically to complement and contrast with a series of ancient stone bridges further along the route
Photo A: © Conzett, Bronzini Gartmann AG
Photo B: © J. Anglade

Figure 2.10 This block laminated footbridge forms part of the renovations at a paper mill near Steyermühl, where a museum of the industry has been installed alongside continuing modern production. The owners, UPM Kymene Austria, stipulated a timber structure, although the preliminary design involved round, hollow section steel tubing as well as timber. This was dropped in favour of the pure timber solution using this innovative technique that permits very shallow depths of section, leading to a light appearance
Photos © Ingenieurbüro Miebach

Figure 2.11 The under-strutted frame used in steep terrain for an open bridge carrying light vehicular traffic. To provide a durable wearing surface resisting concentrated axle loads, a reinforced concrete modular deck has been keyed to the main frames of the glulam. Note also that for inspection purposes, a permanent galvanised metal ladder has been incorporated in the bracing
Photo © TRADA Technology

2.2.6 Contemporary under-strutted frames

In the correct circumstances, these remain a good solution (*Figure 2.11*). They suit an environment in which this form is part of the vernacular tradition, or is at least aesthetically appropriate. Banks need to be high enough to ensure that flood damage will not occur, so the strutted frame is a form eminently suitable to cross steep rocky gorges.

2.3 Cantilevers

Timber bridge examples in authenticated records date to as long ago as 600 BC, and it is suspected that even before this, ancient cultures – including those in China, the Persian Empire and around the Mediterranean rim – had sophisticated structures in this material. In the early days of photography, travellers recorded timber bridges in the Himalaya that belong to the traditions of former millennia. *Figure 2.12A* shows a Bhutanese cantilever bridge whose beams can be withdrawn from their sockets at times of siege. On both sides of the mighty peaks, ancient iron-chained suspension bridges with timber decks also remain in daily service.

Figure 2.12 A: covered timber cantilever bridge leading to the Paro Dzong, Bhutan, 130 foot span, photographed by the author in 2004. This same structure was also photographed by John Claude White and published in London in 1926. It probably dates from the 17th century.
B: covered cantilevered beam bridge at Chengyangqiao, Guangxi, China.
C and D: Kai-no-Saru-bashi Bridge ('Monkey Bridge'), Ohtuki, Yamamashi Prefecture, Japan; carefully restored using tiles to protect the historic cantilevers
Photo A: © CJM
Photo B: © Wikimedia Commons
Photos C–D: © Prof. Honda

In the absence of efficient end jointing, cantilevering is almost an instinctive method of reaching forward with the span and extending the opportunity afforded by limited lengths of material. A famous painting by Carpaccio c.1494 shows a lifting bridge in the naval dockyard in Venice, which was built in this manner.

2.3.1 Modern cantilevers

These are less common, but interesting examples such as the Glennerbrücke, Peiden Bad (*Figure 2.13A*), are occasionally found. These interlaced, cantilevered frames are an outstanding modern interpretation of this ancient system – compare it with the photograph of the Bhutanese bridge at Paro (*Figure 2.13B*).

2.4 Suspension bridges

2.4.1 Historic types

Early suspension bridges used timber walkways, along with other natural materials – such as twisted cane and rattan – for the cables and supports. In other cases, there *were no* walkways, the user simply balanced on the ties, sometimes assisted by a higher-level tightrope, or both people and domestic animals travelled perilously in a woven basket suspended from a slider cable. Where gorges and deep river valleys were crossed, the anchorage points relied on natural features such as trees or rocks. Needham cites an eye-witness report of these from a traveller on the Sichuan–Xizang border in the 17th century. Knapp illustrates Anlan Bridge, Guanxian, Sichuan, a crossing of more than 300 m. Originally this had nine tension spans of plaited bamboo, in use since the third century BC. It is now restored using steel cables, but with its stone piers and pagodas intact. In many remote regions of Asia and the South Pacific, large numbers of similar bridges remain in use.

According to DeLony, wrought-iron chain suspension bridges in conjunction with timber decks (*Figure 2.14*), are recorded in China during the Han Dynasty, c.206 BC. Needham also gives details of the early invention of the metallurgical process by the Chinese. In the early 20th century, photographs

Figure 2.13 A: the interlaced, cantilevered frames of the Glennerbrücke, Peiden Bad, by Conzett, Bronzini, a modern interpretation of an ancient system. Compare with details of a Bhutanese cantilever bridge (B)
Photo A: © Stijn Rolies
Photo B: © CJM

Figure 2.14 Suspension bridges using wrought iron chains and timber board decks;
A: Bhutan, at the confluence of two rivers eventually joining the Brahmaputra;
B: Luding Bridge, Sichuan – in 1935, this was the site of a famous victory of the Red Army; currently restored, photograph showing work on an entry pagoda
Photo A: © CJM
Photo B: © Wikipedia GNU

Figure 2.15 After a few months of rigorous site work establishing the reinforced concrete tie-back foundations, retaining walls, and launch platforms, the second Traversina Footbridge was completed in August 2005, following an erection process that owed more to mountaineering than conventional civil engineering. A pair of main steel cables defines the high-level tilted cantenary shape above the ascending timber staircase. From the high wires descends a cross-laced network for the foot-way suspension. All is held together using cable clamp fixings. Mounted longitudinally on cross-beams are ten parallel glulam timber flexural-tension members, in two sets, forming the lower chords
Photo © Hugh Mansfield-Williams

by Forrest, the Scottish plant collector, showed similar suspension bridges in Yunnan, China. Luding Bridge, Sichuan, is another fine example, now being restored as a symbolic site of an episode in 'The Long March'. The exact nature of this battle is disputed, since – as with many aspects of Chinese history – there are accusations of exaggeration countered by eye-witness accounts supporting the orthodox version. But it seems likely that a certain number of Red Army soldiers did actually scramble along the chains, which had been maintained by a family of blacksmiths, and who had done so for centuries. Monastic wall paintings in Bhutan, associated with Thangtong Gyalpo, the mystic bridge builder, show that these examples are very similar to centuries-old designs on the other side of the Himalaya.

The introduction of steel cabling permitted impressive suspension bridges with very long spans. Cabled lateral stiffening trusses were introduced, along with selected, regularised decking in timber. With European support, modern steel and timber suspension bridges, for which a series of design manuals is available, have been constructed in Nepal. These are ideal for the mountainous terrain, where transport largely remains dependent on pack animals and small tractors.

2.4.2 Contemporary forms
Figures 2.15 and *2.16* show one of the latest imaginative timber structures designed by the engineer-architect Jürg Conzett, of Chur, Switzerland. Earlier, an avalanche destroyed another impressive bridge at this site. This first structure used an inverted bowstring frame, strutted beneath a lightweight timber deck.

As we shall see in Chapter 5, the classic suspension bridge, supported from masts or towers, is also seen in modern timber. Pure tension-ribbons have been built recently, spanning from high banks in the case of the Main–Donau Canal bridge, or carried on tall braced towers and trestles, as is the case with the Drachenschwanz at Ronneburg (*Figure 1.8*).

2.5 Arches
Both in the ancient and modern world, some of the most notable bridges are arched, and those in timber are no exception.

2.5.1 Roman arches
Apollodorus of Damascus was the chief military engineer for the Roman Emperor Trajan (AD 98–117). The emperor directed a campaign in Dacia, modern-day Romania, for which a famous and long (approx. 1200 m) multiply arched permanent timber bridge was built (*Figure 2.17*). Leading to the modern-day Serbian river banks, for more than 1000 years this was

Figure 2.16 The gorge drops a further 70 m below the deck, and while these out-lying timbers are slightly spaced apart for ventilation purposes, they seem 'solid' enough to screen the view downwards, reducing sensations of vertigo. The cable-suspended footbridge spans a gap of 56 m, with a deck length of 62 m and an ascent of 22 m. Access to the site is only possible along a walking trail, so all of the timberwork had to be transported in, and was consequently prefabricated as much as possible. Also featuring strongly in the design work was the need to avoid flat upper surfaces and to provide only small contact locations between elements, with the circulation of free airflow maximised. In all of the timber elements, only heartwood was permitted, using larch and fir. Hot-dip galvanizing treatments were applied to all exposed steelwork
Photo © Hugh Mansfield-Williams

the longest arch bridge ever constructed. It carried the Via Flamina, a road that began in Northern Italy and passed the shores of the Adriatic, crossing the Alps.

After approaching Vienna, the road dropped into a region where the 800 m wide Danube has rocky shores and waters more than 50 m deep. Near the cities of Drobeta-Turnu Severin (Romania) and Kladovo (Serbia), pontoon bridges were soon consolidated by constructing the permanent timber bridge. Apollodorus recorded its details, and his documents survived until at least the 12th century, during which time they were cited by several secondary authors. Unfortunately, his records are now lost, but the bridge is shown in detail on Trajan's Column in Rome, and on commemorative bronze coins issued for the opening in AD 105. Since Apollodorus also directed the carving and erection of Trajan's Column, it is likely that these representations are very accurate.

The timberwork spanned between 20 intermediate masonry piers onto abutments that were under-pinned with wooden caissons and timber piles, a common Roman technique. The abutments and 12 piers still stand, and these were archaeologically documented in 1982.

Figure 2.17 Apollodorus' Bridge, modelled 1 : 50 from details on Trajan's Column, by Giuseppe Paolillo, University of Florence. Each structural span is estimated to have been about 33 m. Six arched frames were arranged parallel to one another in plan. Each individual arch-frame comprised three sets of thick, bent and spaced laminations. These supported a high-level timber deck, together with triangulated struts arising at the piers. Typical Roman bracing was of saltire form, as indicated by the relief images on Trajan's Column
Photos © CJM

2.5.2 Timber arches in the Far East

Chinese timber 'rainbow bridges' date from the Song Dynasty. A famous scroll painting from the era, executed in ink and coloured pigments, is popularly known as 'Along the River during Quing Ming Festival'. Measuring 25.5 cm high by 525 cm long, the scroll was painted by Zhang Zeduan, a highly regarded court painter, who depicted myriad details of Chinese life and technological practices during his lifetime (AD 1085–1145). Shen and Liu describe the special form of such structures, which are constructed by 'weaving' straight round timbers, crossing these over and under transverse square baulks. As an exercise in experimental archaeology, a full-sized re-build of the Quing Ming Festival Bridge was constructed at Jinze, west of Shanghai, in 1999. Bibliographic research

Figure 2.18 Kintai-kyo, also known in English as the 'Bridge of the Brocade Sash', in Iwakuni, Yamaguchi Prefecture, Japan.
Photos © Tomo.Yun (www.yunphoto.net/en)

also discovered 13 similar bridges within the original region and date. Subsequently, physical explorations have revealed numerous impressive and still-standing ancient Chinese covered, cantilevered and arched bridges, including about 100 of these in Fujian Province alone.

In Iwakuni, Yamaguchi Prefecture, Japan, the Kintai-kyo Bridge (*Figure 2.18*), also known in English as the 'Bridge of the Brocade Sash', was founded in 1673. This bridge gave the title to a book by Sachaverell Sitwell recording his travels and observations in post-war Japan. It reflects a distinctly individualistic approach to conservation, which is also traditional for temples and pagodas in Japan. It has undergone regular re-building span by span. In previous centuries this was performed every decade, by imperial decree. The most recent reconstruction followed floods and a typhoon in 1950. Compare the method of construction with the cantilevers in *Figure 2.13* – the principle is now extended into a pure arch.

2.5.3 English historic arches

The Canaletto scene of 'Old Walton Bridge over the Thames', now to be seen in the Dulwich picture gallery, was painted in 1754 for the artist's patron, Thomas Hollis, the local MP. It is a good example of the triangulated and latticed polygonal approach to timber arches, very much in vogue at the time, and also exemplified by the 'Mathematical' bridge of Queens' College, Cambridge, and a smaller surviving example in West Wycombe Park, Buckinghamshire. Although not following the same structural principles, the 'Chinese Bridge' at Godmanchester, Cambridge (*Figure 2.19B*) is aesthetically similar. Early in the 20th century a fine mechanically laminated Greenheart bridge with twin skewed elliptical arches was completed on the River Weaver, near Northwich (*Figure 2.19A*). Fortunately this unique 'listed building' has been accurately restored; some of the work undertaken is described in more detail in Chapter 8.

2.5.4 Glulam arches

As a contemporary timber engineering material for bridges, glued laminated timber (glulam) is discussed in Chapter 4. The majority of the modern projects throughout this book are constructed with one form or another of this material. Developments in the use of glued – as opposed to mechanically laminated – timber began earlier than most people realise. In around 1807–1809, a Bavarian engineer named Carl Friedrich Wiebeking developed horizontally laminated timber arch bridges with spans of up to 60 m. Booth, in the reference cited above (note 9), discusses these. Most of them used thick iron-bolted or rod-connected laminations of oak. But because of limitations with the technology, the first glulam bridge at Altenmarkt used spruce, which was easier to bend than oak. It was fabricated in situ, working (presumably with great difficulty) from scaffolding. Probably, this choice of timber, together with the lack of understanding of good protective design detailing and poor adhesive durability, led to its short life.

By the start of the 20th century, patents were being granted for glulam in Germany. Within the reference given, examples are provided of early 20th-century structures laminated with casein adhesive in Switzerland, which still stand today. Provided the material bonded with this adhesive is kept dry, it has substantial longevity. In 1939, in the United States, a landmark technical publication appeared that strongly influenced subsequent North American codes. This was entitled 'The glued laminated

A

B

Figure 2.19 Historic British timber arches. A: Dutton Lower Horse Bridge, on the River Weaver Navigation, Northwich, Cheshire; John Saner, a waterways engineer, designed this elegant listed structure in 1910 – it is a unique example of doubled skewed arches from mechanically laminated greenheart.
B: the 'Chinese Bridge', Godmanchester, Cambridgeshire, Grade-II Listed, 1827, photographed before the most recent restoration in 1993
Photos © CJM

wooden arch'. By then, glulam bridges were well established. Several of the 'Hetzer system' bridges in Europe are illustrated within the report, including those designed by Terner and Chopart in Lausanne, 1910, and another in Copenhagen, 1929, which carried motorised traffic.

2.5.5 Contemporary arches

Highly developed and competitive timber arch bridges are in use in the 21st century (*Figures 2.20* and *2.21*). Glulam has several advantages for these, primarily in that it is available in large cross-sections and lengths that can be formed into curves. Its production is carried out at a closely controlled moisture content, and components are often installed at a lower value than the eventual mean equilibrium moisture content of the environment. This is because, generally speaking, swelling due to moisture uptake is easier to provide for than shrinkage due to uncontrolled drying.

The new 70 m main span Tynset Bridge, completed in May 2001, typifies the highly developed and competitive timber structures that are provided in the 21st century. For the shorter 27 m spans, two pairs of three-hinged arches are used, together with a record-breaking 70 m latticed-arch main span. These give a total bridge length of 125 m. The decks are of stress laminated timber, supported by steel crossbeams and suspended from the arches by steel tension ties. The river crossing over the Glåma in Hedmark County, Norway, is approximately 400 km north of Oslo. The bridge was designed to serve the local community and to provide access for regional traffic for 100 years, with full traffic loads. It includes two traffic lanes and a 3 m pedestrian walkway. The project challenges included: selecting the optimum timber structure for the site, solving technical issues associated with the connections, and finding technical solutions to ensure a 100-year design life for the structure. For the structural engineers, the bracing posed a problem which was resolved by thickening the lower segments so that their lateral stiffness was adequate

Figure 2.20 Ollas Overpass, Espoo, Finland – a pedestrian and cycle road-crossing bridge. The static system uses two-hinged arches, located each side of the deck. Neoprene bearings support the timberwork and where it runs off the arches, transoms and triangulation brace the stressed laminated deck
Photos © Puu/Finnish Timber Council

2.6 Trusses

As shown in the introduction to this chapter, trussed structures are the most advanced of the five primary forms, in the sense that for successful design, they require a means of achieving effective node connections. Roman bridge types contained triangulated linear forms, over which experts argue as to whether or not 'true truss action' was achieved. However, by the end of the 18th century, the structural mechanics of the truss were widely understood, while as early as 1579 Andrea Palladio was responsible for propagating the same principles (*Figure 2.22*). The swift rivers of Northern Italy regularly swept away designs whose supports were piled into the riverbed, so Palladio required clear spans and, as discussed by Heyman, he showed how these could be achieved through triangulation and metal-connected nodes.

Post-industrial revolution forms sometimes exhibited transitions – for instance, from fans to truss types. *Figure 2.23* shows two further examples of interpretations of historic truss forms using modern timber engineering technology. The 'fishbelly' design is appealing both to the engineer and to the lay person who instinctively appreciates its efficiency. However, for security of the structure, it requires a substantial clearance below its lower chords.

Figure 2.21 Tynset Bridge
Photo © A. Lawrence

Figure 2.22 The first and second of Andrea Palladio's four 'Invenzioni' were trussed footbridges that are essentially statically determinate; his ideas were remarkably advanced for the time (1579), recognising the importance of achieving clear spans, and using iron straps and bolts at the nodes. It is known that one of these was actually built and it is believed to be that shown in B – spanning 100 ft over the river at Cismone. Palladio's treatise also contains an arch, while his simpler trestle bridge at Bassano still stands. Models – University of Florence

Photos © CJM

2.6.1 North American patented trusses

In the early 19th century in North America, there was a proliferation of patented timber bridge designs. Nowadays, these are well documented by government agencies and national covered bridge societies, who have introduced a cataloguing system, forming an effective pressure group for conservation. The complete range is too varied to describe here, but explanations of all the types and variations, actual construction and rehabilitation examples, together with their loads, structural analysis, connections and design issues, are available in a recent and comprehensive *Federal Highway Administration Report*.

Amongst the most important of these forms are those using alternatively the Town or the Howe types of truss (*Figure 2.24*). Such bridges are of immense historic importance, and for this reason they are often protected by incorporation in the USA National Register of Historic Places. These particular types are singled out for illustration and further brief discussion, since, like the Burr Arch Trusses described in Section 2.7 (where we deal with hybrid forms), they include useful concepts still adopted by contemporary designers.

Figure 2.23 Two examples of interpretations of historic truss forms using modern timber engineering technology. A: three-pinned open trussed girder footbridge of 41 m span located between Adliswil and Wollishofen, Switzerland – a modern interpretation of a 1935 design, not dissimilar to the Palladio trussed bridges shown in *Figure 2.22*.

B: At Walshausen Manor, near Hildesheim, an older bridge of this same lenticular shape was recently replaced. This is known as 'Lavesbridge' after the distinguished German architect and engineer Georg Ludwig Laves (1788–1864), who designed both the manor and the 'fishbelly' bridge. He worked mainly around Hannover. The older bridge was swept away by floods in 1947, and although attractive, this form is clearly rather susceptible to this risk. (Mariac Footbridge). Note also the protective metal covers on the top chords of both booms

Photo A: © CJM

Photo B: © French Wiki

In 1820 Ithiel Town of Connecticut patented the latticed all-timber trusses, while in 1840 William Howe of Massachusetts combined timber with adjustable wrought-iron tie rods. In the latter case, the screw-threaded ends of the vertical metal ties provide the advantage of simple adjustment during construction and ease of subsequent re-tightening. However, the simpler Town truss put the capability of bridge building within the hands of most small-town and village carpenters. Town wished this to happen, aiming to earn income not by building them himself, but by selling his standard designs and charging royalties on the completed structures. This was a successful venture, making him very wealthy. He advertised the designs with the catchy line that 'they could be built by the mile and cut off by the yard'! It it also worth noting that both of these forms are extensively prefabricated and modularly arranged for simplicity of transport and erection – an important modern principle.

In the 19th century, both of these types were introduced into Europe. They are still used for contemporary structures on both sides of the Atlantic. Of course, in the case of the Howe type, modern steelwork replaces the adjustable iron ties, and in both instances modern timber engineering techniques are applied to the connections, with materials such as glulam often used for the members.

The Howe truss made its way to Switzerland at an early date. In the home of the timber bridge, examples include the Obermatt Bridge, in Berne canton, built in 1903 with a 32 m span, and still in use, standing on the boundary between Langnau and Lauperswil. Also, on the Schwarzenburg-to-Freiburg road, the Sodbachbrücke is a still-older example of a 'Swiss Howe', constructed in 1867, only 27 years after the original patent. In 2000, in Böheimkirchen, Lower Austria this historic system was revived for an attractive newly built pedestrian and cyclist bridge using glulam and other contemporary timber engineering techniques.

There is also a significant number of Howe types in Canada, and an impressive example is the Hartland Bridge, crossing the Saint John River at the town of Hartland in New Brunswick (*Figure 2.25*). With seven spans formed from Howe trusses, this is believed to be the world's longest covered bridge. Constructed in 1901, the roof was not added until just after 1920, when repairs were made following serious damage by river ice. Collisions by vehicles ignoring clearance restrictions are a hazard to covered bridges, and for this reason several subsequent phases of repairs have occurred. Most recently, structurally unsound members were repaired and some replaced commencing in 2006, with the bridge re-opened to light traffic – and with very strict access controls – in the summer of 2009.

Figure 2.24 A: Waterford covered bridge, Pennsylvania, an excellent example of a Town lattice truss bridge; built in 1875 and restored in 2001, with light wheeled traffic still permitted. Now a bridge on the US National Register of Historic Places.
B: Jay Bridge in New York State, exemplifying the Howe truss type. Compared with the Town truss – although this also comprises a diamond lattice – the triangulation is much larger as are the internal web cross-sections, and the adjustable vertical metal tie rods can be clearly seen
Photo A: © Dtbohrer
Photo B: © Mwanner

Figure 2.25 Hartland Bridge – crossing the Saint John River at Hartland, New Brunswick – seven spans of Howe trusses. At 391 m, this is the longest covered bridge in the world
Photo © Wikimedia Commons

Figure 2.26 Vihantasalmi Bridge, Finland – total length of 180 m, with three main spans of 42 m each as glulam kingpost trusses – heavily protected and involving a composite timber–concrete–steel deck; for Highway No. 5 crossing the Vihantasalmi strait 180 km north of Helsinki; a prize-winning design completed in 1999
Photos © CJM

Reiterating the fact that contemporary benefits are sometimes gained from these designs, the *US FHA Covered Bridge Manual* describes and illustrates a project carried out in 1992, to build a new Town lattice truss bridge. This 44 m span structure with 4.6 m high trusses crosses the Speed River, in Guelph, Ontario. The City Parks department commissioned this Douglas fir 'trail bridge', with the planning, carpentry and erection labour all supplied by members of the Timber Framers Guild of North America.

2.6.2 Contemporary trusses

Trusses remain competitive in minimising material consumption, and providing lightweight components that are readily prefabricated and ready to erect. Projects such as the Flisa Bridge (Chapter 9) and Vihantasalmi Bridge (*Figure 2.26*), show remarkably good cost comparisons with steel, even throughout the cyclic commodity price fluctuations of both materials.

Figure 2.27 Contemporary timber trussed bridges are built both as open and covered types.
A: Beston Bridge, Norway, 24 m span, part of the Oslofjord Crossing system, designed to carry local traffic over a national highway.
B: Andelfingen, Switzerland, an interpretation of the Howe truss using glulam and high-strength protected steel rods
Photos © CJM

Contemporary timber trussed bridges are built both as open and covered types (*Figure 2.2*). The recent Nordic timber bridge initiatives have led to well-founded environmental index studies. Using relatively low volumes and greater proportions of bioenergy-based materials, manufacturing the timber bridge consumes roughly two-thirds of the energy of the best alternative, comparing steel concrete and timber, while the total emission of three greenhouse gases is about half of that for the alternative types. Transportation factors are also improved through the use of relatively closely available resources and lower masses.

Figure 2.28 Elevation of Grubenmann's Schaffhausen bridge, engraved in 1799, the year of its destruction during the Napoleonic wars. The Bishop of Derry hoped Grubenmann would build a 600-foot span version over the river Shannon!
Drawing © Sir John Soane Museum

2.7 Hybrids

By the 18th century, impressively long timber spans were being achieved. Often these involved hybrid trussed-arched frames, although over the river Limmat at Wettingen, an exceptionally modern pure laminated arched bridge was also built in around 1795 by a group of talented carpenters. Typical European trussed-arch examples include a Rhine bridge, constructed at Schaffhausen in 1758 by Hans Ulrich Grubenmann (*Figure 2.28*).

Schaffhausen Bridge had an overall length of approximately 380 feet (the larger span being 193 feet or 58.8 m), with the construction landing on an allegedly redundant pier at mid-span. A story that is probably apocryphal is that the cautious Bürgermeisters deterred the proud carpenter from crossing the river with a single span. The structure had laminated arched ribs, each with a depth of about 2 m and comprised seven courses of timber, notched and banded together. Today, the Hasle-Rüegsau bridge, below Burgdorf, in Canton Berne, can be considered a direct descendant of the Grubenmann hybrids. Containing a pair of 60 m span trussed arches, mechanically laminated, this remains available for light traffic of up to 3.5 tonnes, after a life of nearly 170 years.

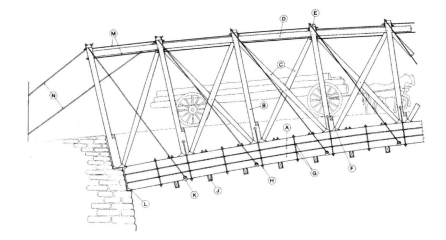

Figure 2.29 Early 19th-century growth in North America gave rise to some very large timber trussed-arched spans, one of the most noteworthy being the 'Colossus Bridge' over the Schuylkill river at Philadelphia, USA. This was constructed in 1812 by Lewis Wernwag, and had an amazing free span of 340 feet (102 m). The laminated arch elements each comprised six 6 × 14 inch (150 × 350 mm) heart-sawn baulks of white pine. For durability, these were separated, but linked, using iron bands and threaded rods
Drawing © Lee Nelson
Painting © American Society of Civil Engineers

At this point it is convenient to consider developments of the important hybrid, the trussed-arch, in a different part of the world.

In North America, the Burr arch truss, patented by Theodore Burr in 1817, evolved into a very successful proprietary type (*Figure 2.30*). Evidently the combination of the arch with the truss was empirically successful, despite it frustrating various later attempts at explanation by computer modellers. Once pressure-preservative-treated timbers started to become regularly available, bridge builders were encouraged, perhaps rather rashly, to attempt similar structures without a roof. In this manner, an amazing example of nine spans of double-arched truss-arch hybrids was erected over the estuary at Fredericton, New Brunswick (*Figure 2.31*). Unroofed Burr-type bridges and similar forms continued to be built in Canada right up to the 1940s, and a few survivors of this era remain today.

2.7.1 Modern hybrids
Provided they are achieved with elegance and have some purpose in their form selection, there is still a place for hybrid solutions. *Figure 2.32* shows an under-strutted frame in which the radiating compression members extend to positions beneath the roof, where some of the bracing is concealed. Hangers support the deck. This is a good example of a modern hybrid form – neither pure beams nor exactly arches.

Chapter 5 includes a short case study of the Passerelle de Vaires-sur-Marne, a French railway crossing for passengers that adopts some of the concepts of the Burr arch truss.

2.7.2 Moving bridges
These are often hybrid forms and there are several practical moveable bridge arrangements in timber. These include the classic bascule bridge (*Figure 2.33*), familiar in the Netherlands and at Arles in the South of France. Both single and double leaf forms are possible. On inland waterways, swing and tipping bridges in timber were once common and

Figure 2.30 Covered Burr arch-truss bridges in North America;
A: Bridgeport, California, founded in 1862, one of the longest single spans.
B: interior view of Baumgardener's Bridge, 1869, Pennsylvania – clearly showing the sharing of structural duty between the arches and the trusses, the latter supporting the roof and the former being laterally braced.
C: Philippi Covered Bridge, West Virginia, USA; built in 1852, and restored several times, it has double spans, each approximately 150 feet (45 m) using two sets of Burr arch trusses; with a width of 26 feet (7.8 m), it is one of the few surviving 'double-barrelled' (two-lane) examples; with strong American Civil War associations, it was constructed by Lemuel Chenoweth; now it is unique in another way, being the only remaining covered timber bridge on the US Federal Highways
Photos © Valerius Tygart

Figure 2.31 Nine spans of double-arched truss-arch hybrids were erected over the estuary at Fredericton, New Brunswick. The bridge was eventually destroyed by ice floes
Photo © NB Archive

there is still a demand to restore or revive them. Moving bridges that tilt, and types that lift from both ends or can sink below the navigation level (using dense timber such as Greenheart) are further possibilities.

2.7.3 Modern versions

It is of practical importance in docklands, harbours and inland waterways to provide clearance for waterway traffic. Many regeneration and refurbishment schemes have been undertaken in such areas, continuing the need for walking and cycle routes, and the linking-up of riverside areas. Modern timber design methods, materials and fabrication concepts can provide similar solutions to those used in the past for industrial duties. Most of the types of deck discussed in Chapter 6 can be contemplated for moving bridges. All of the material types discussed in Chapter 4 can be considered, including glulam, the recently introduced block laminated timber and the much older mechanically laminated timber. The facility to taper glulam helps to lower the weight of the lifting sections and the balance beams (upper members that support counterweights) can also conveniently be laminated. Both traditional and contemporary architectural styles are possible, but the spans for this type of bridge tend to be fairly modest, with those in excess of about 24 m being rare.

2.7.4 Future prospects

As we have already seen in Chapter 1, innovations continue, providing prospects for a bright future. The sustainability credentials and fine appearance of timber bridges commends them to prospective owners. In Chapter 3 we will see how durability has finally been positively secured, and *Figure 2.34* reminds us that an inspired design always requires practical measures to ensure longevity. Examples of structural timber sensitively combined with steel, for example in the Almere Pylon Bridge (Chapter 9), show how aesthetics can be matched to contemporary sensitivities.

Figure 2.32 A historic Swiss type observed by John Soane in his 18th-century travels. Here a new bridge crosses the River Töss at Wülflingen, with a 47 m span, for pedestrians and cyclists
Photo © CJM

Figure 2.33 The lift bridge is a traditional moving type. In this small, modern interpretation – 'Ontario Bridge' at Salford Quays – the mechanism to raise the decks has been hidden within hollow glulam posts. Fabrication by Cowley Timberwork
Photo © CJM

Figure 2.34 A contemporary interpretation of a design for a much larger masonry bridge by Leonardo da Vinci – his was intended to cross the Bosphorus; this version provides a pedestrian crossing in Norway. Originally, the relatively flat upper surfaces of the splayed laminated arches were protected only by surface treatment, but soon covers of the types discussed in Chapter 3 were added. An inspired design always requires practical measures to ensure longevity
Photo © CJM

3 Durability and protection by design

3.1 Introduction

Timber bridges can certainly be designed to retain their original performance, without undue deterioration through lack of durability. Natural organisms have evolved in such a way that, left to its own devices, a fallen tree in the woods or forest will decay – quite rapidly, or very gradually, dependent upon factors including the botanical species and associated natural durability of its heartwood. But no construction material is capable of lasting forever, and as evidenced by the life span of the older timber bridges discussed in Chapters 2 and 8, those built of timber will achieve long and excellent design working lives, provided that the correct principles are observed.

Structures must be protected from moisture, since the natural decay organisms mentioned above require the presence of water and warmth to operate. In 'protection by design' (*Figure 3.1*) the owner's requirements are matched with available information on the durability of the materials and the recommended protection methods and details. Two options are

Figure 3.1 Alternative protection strategies.
A: fully roofed bridges are still appropriate under some circumstances – Leimbach foot and cycle bridge, Switzerland.
B: another option is to introduce a series of 'protection by design' arrangements – in this case, double pressure preservative treatments and local non-corrosive metallic covers over key elements – Mølledammen Road Bridge, Norway.
C: Agents and their principal effects, and protection measures against these effects
Photos © CJM

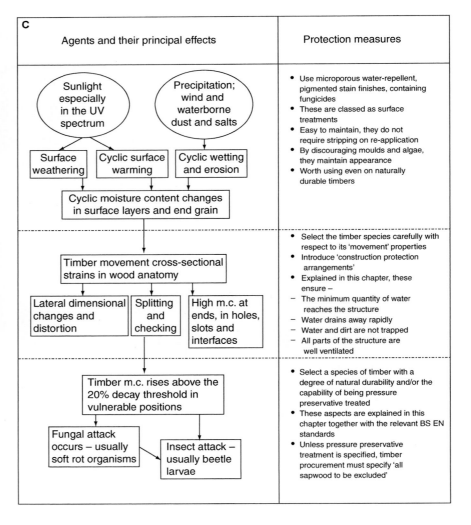

Agents and their principal effects	Protection measures
Sunlight especially in the UV spectrum / Precipitation; wind and waterborne dust and salts → Surface weathering / Cyclic surface warming / Cyclic wetting and erosion → Cyclic moisture content changes in surface layers and end grain	• Use microporous water-repellent, pigmented stain finishes, containing fungicides • These are classed as surface treatments • Easy to maintain, they do not require stripping on re-application • By discouraging moulds and algae, they maintain appearance • Worth using even on naturally durable timbers
Timber movement cross-sectional strains in wood anatomy → Lateral dimensional changes and distortion / Splitting and checking / High m.c. at ends, in holes, slots and interfaces	• Select the timber species carefully with respect to its 'movement' properties • Introduce 'construction protection arrangements' • Explained in this chapter, these ensure – – The minimum quantity of water reaches the structure – Water drains away rapidly – Water and dirt are not trapped – All parts of the structure are well ventilated
Timber m.c. rises above the 20% decay threshold in vulnerable positions → Fungal attack occurs – usually soft rot organisms / Insect attack – usually beetle larvae	• Select a species of timber with a degree of natural durability and/or the capability of being pressure preservative treated • These aspects are explained in this chapter together with the relevant BS EN standards • Unless pressure preservative treatment is specified, timber procurement must specify 'all sapwood to be excluded'

illustrated above, and a third, explained in more detail below, is to build only with timbers that are classified as naturally 'very durable'. But all of these judgements need to take into account aspects such as the site and its exposure conditions, the expected quality of workmanship and the anticipated maintenance. Not all of the parts of a bridge need to be designed so as to achieve the design working life applicable to the principal structure, especially if accessibility for maintenance and ease of replacement have been taken into account.

The historic approach to ensuring longevity was to provide a complete roof, and sometimes this may still be appropriate (*Figure 3.1A*). There are certainly suitable materials and timber engineering techniques to accomplish this successfully. Some modern structural panel materials are discussed in Chapter 4. Another well-proven protective design practice (*Figure 3.1B*) is to use ventilated local covers, sometimes of corrosion-resistant metal on an easily replaceable timber base, to achieve similar sheltering. This will be examined in detail in this chapter since it is now a popular and successful solution. In this general technique, various wood-based screening materials are also employed. In all cases, the objective is to reduce or eliminate standing water on the key structural elements.

Note that in both of the examples shown in *Figure 3.1*, at least two 'layers' of protection are included. In the case of the covered bridge, not only is there a modern high-quality waterproofed roof, with plenty of overhang in the eaves, but water-repellent stain finishes are applied to all of the timber parts; rain screening is attached to the sides of the stress laminated deck, and the tie bars are protected. For the uncovered through-arch road bridge (*Figure 3.1B*), local non-corrosive metallic covers are placed over the key elements and a double process of pressure preservative treatment has been applied to the principal structure and the deck. Timber bridge designers nowadays regard this 'belt and braces' approach as essential.

In the early days of 'modern' timber engineering, during the first half of the 20th century, the means towards longevity was to rely heavily on pressure treatments. This is still not entirely inappropriate, but there is a tendency now to wish to avoid the use of synthetic chemicals where possible. This leads to the third string in the bow, which is detailed design measures to accomplish water-shedding and ventilation, to maximise cleanliness of the structure and generally to provide as much protection as possible. Good timber bridge design entails a successful blend of these three approaches, as well as taking maximum possible advantage of the natural durability of the heartwood of selected species of timber.

3.2 Design working life

The definitions of 'design life' are not entirely uniform, although nearly all approving agencies and codes of practice make reference to it. Throughout the Structural Eurocodes, the term 'design working life' is used, and BS EN 1991-1 defines this as:

> The assumed period for which a structure or part of it is to be used for its intended purpose with anticipated maintenance but without major substantial repair being necessary.

Table 3.1 Design working life classifications

Class	Working life (years)	Examples of appropriate structures/components
1	10	Replaceable parts of small, low-cost footbridges, e.g. open decking, handrails
2	25	Replaceable non-load-bearing parts and components of normal-quality footbridges, joinery items, e.g. parapet rails, battens and spindles. Items subjected to wear, scheduled to be replaced during normal maintenance – e.g. steps and running boards
3	50	Larger components of pedestrian and light motorised-traffic bridges, including accessible and removable load-bearing items
4	100–120 approximately	Load-carrying and principal components of large motorised-traffic bridges; listed bridge structures as the design working life target for repairs

In this chapter, we shall see how the phrase 'or part of it' is important.

Table 3.1 indicates classifications of design working lives for components and complete bridges. Alternative periods may be chosen, in accordance with the owner's wishes. The indicated period of 120 years for 'bridges' in the National Annex to BS EN 1991-1 relates to major highway or railway bridges.

Timber road bridges are indeed designed to survive for this period, good examples being the large Norwegian road bridges included in the case studies in Chapter 9. However, it should be questioned whether this target, with an implied additional capital expenditure, is always appropriate for smaller footbridge and light traffic bridges that would usually be more sensibly designed for a target of 50 years.

Generic definitions relating to durability are also indicated in BS 7543, principally for buildings, but also applicable to bridges. 'Service life' therein is defined as 'the actual period of time during which no excessive expenditure is required on operation, maintenance or repair of a component or construction'. An 'agent' is 'whatever acts upon a structure or its parts, e.g. sunlight, wetting, structural actions including loads and temperature effects'.

Figure 3.2 Examples of protection classes.
A – totally protected;
B – partly protected;
C – unprotected. Using 'very durable' species. Switzerland; France; England
Photos © CJM

3.3 Agents and their effects

The remainder of this chapter is concerned with the strongly recommended construction protection measures that are necessary to protect the bridge against the inevitable agents that are introduced through its normal working life environment. A summary of these agents, their effects and the corresponding protection measures is given in *Figure 3.2*C.

It can be seen that the figure is divided vertically into three main phases that indicate the general progression of deterioration that is significantly slowed or prevented altogether through the measures indicated on the right. Later in the chapter, we shall examine more closely how risks are evaluated and decisions taken on matters such as the choice of naturally durable or treatable wood species. Also, construction protection arrangements are set out in considerable detail, together with illustrations of typical interpretations.

3.4 Protection classifications

Timber bridges and their components can be separated into 'protected' and 'unprotected' items. The logic is to assess the potential durability of the components and compare different options, as well as evaluating the effectiveness of protective construction details.

In these classifications, a difficulty is overcoming the difference between a client's preliminary design aspirations regarding durability, and their notion of realistic budgets. Early in planning, it is essential for construction professionals to discuss the levels of protection and the types of construction advised. Sengler has estimated the additional costs for effective protection to be about 5% for 'protected' timber bridge construction, and 10–15% for 'partially protected' structures. Considering the investment that these figures represent in terms of longevity, and freedom from major repairs, they should be regarded as essential, rather than optional and expensive additions.

Older publications usually describe merely 'covered' or 'open' bridges, but this does not allow sufficiently precise judgements about the quality of protection in the design. For a better indication of likely durability and life expectancy, a classification system for protection measures has been developed in Germany, and similar concepts are now widely respected throughout Europe. The classification approach shown in *Table 3.2* permits better forecasts of durability, susceptibility to damage, and likely maintenance costs. Because of the necessity for experience-based judgements, a refinement has been made by distinguishing between 'totally protected' components, which are never subjected to direct wetting, and 'partly protected' components which may occasionally be exposed to precipitation or wind-blown moisture.

The aim of most modern timber bridge designs should be to achieve at least the 'partly protected' category for the load-bearing elements, as well as for larger components that are expensive and/or difficult to replace. This facilitates the selection of a wider range of timbers and wood-based materials, since the exclusion of timbers other than those qualifying as 'very durable' can restrict choice. For the various bridge components, different criteria may be applied, so that structural elements, large joinery items and protective claddings and roofing can all be assessed separately. Then the materials' inherent durability, potential for treatment and other factors such as appearance and weathering can also be related.

Figure 3.3 Examples of substantial partial protection. A: louvres outside a stressed laminated deck; countersinking of mechanical fasteners; kiln-dried oak covers on metal hangers; microporous water-repellent stain finish; Leimbach Bridge nr. Zurich. B: spaced-ventilated larch louvres on a Swiss under-strutted bridge
Photo A: © CJM
Photo B: © TRADA Technology

Table 3.2 Classification for degrees of protection

Protection class	Examples of appropriate measures
1 Totally protected	Completely covering the elements/components concerned*
2 Partly protected	Protection through which precipitation may filtrate, e.g. louvres, open cladding, metal covers, an open deck geometry that ensures rapid drainage and ventilation
3 Unprotected	Absence of above types of feature, reliance for example upon 'very durable' wood species, and with protective detailing, e.g. prevention of water/dirt trapping – still important!

Note
* Not necessarily with a roof over the whole bridge.

3.5 Fundamental arrangements

The factors that influence the actual service life of any bridge or its main components are:

- original targeted design working life – some historic bridges have lasted much longer than originally intended, others failed prematurely;
- the general surrounding environmental conditions – e.g. inland or coastal;
- the use of the bridge and the nature of its traffic, which may have changed during the lifetime;
- structural detailing and construction protective measures;
- the materials used, and their actual durability;
- the original quality of workmanship;
- the actual maintenance and repair that has occurred.

3.6 Biological agents

The terrain and contours, the below-deck clearances and the nature of the crossing usually limit the choices of bridge form. Nevertheless, it is good advice to try to select a structural system that inherently offers improved prospects of durability. For example, where possible, a deck is always better located so that it protects the main structure. The manner in which the deck itself is protected is another important choice. Once the vital preliminary arrangements have been decided, there is considerable scope in the geometry and shaping of members to shed water, minimise contact between elements, and generally to optimise the structural detailing. There are also choices of specific materials and sometimes their treatment, which are likely to provide enhanced durability.

In north-western Europe, including the whole of the British Isles, by far the worst immediate risk for timber structures is fungal attack. Where insect damage occurs, this usually follows degradation of the material by fungi, in the presence of moisture. Attaining durability is thus strongly dependent on keeping the moisture content below 20%, which is generally recognised as the decay threshold. All forms of mechanical protection such as cladding and full or local metallic covers are aimed at this. Parts of bridges may inevitably have to be located close to the ground or in situations where they are permanently exposed. Here, the options are essentially to use a 'durable' or 'very durable' species from which all the sapwood has been excluded, or to opt for a method of pressure preservative treatment that is recognised as sufficiently robust.

Both fungal and animal (mainly insect) agents of wood degradation belong to wide ranges of biological classification, but their dangers are well understood, and protection systems are well developed. Sunlight and natural radiation, particularly from the ultra-violet spectrum, causes nearly all of the initially new surface shades of natural timber to become silvery-grey. The exact final weathered colour and texture depends on the species, the local exposure situation, including any conditions that encourage surface moulds or lichens, and also the presence or absence of airborne particles, salts, etc. Weathering is also more pronounced if the exposure arrangement leads to occasional or frequent wetting. Precautions against both light exposure and superficial wetting consist mainly of surface treatments. Some pressure preservative systems do include colouring pigments. Also, biocides and fungicides are often blended into surface treatments. However, the principal objectives of these two classes of treatment should not be confused. As explained below, where necessary, only pressure preservative treatment is adequate to significantly upgrade the long-term durability of a species of timber.

3.7 Preservation principles and hazard classes

In avoiding fungal and insect attack, the essential principles are:

- conceive the design so as to eliminate the worst risks and minimise others;
- utilise the natural durability of the heartwood of a wide range of wood species;
- protect the timber structure to the greatest degree possible.

The risk of biological attack is seldom the same in all zones or parts of a bridge, and for this reason, five hazard classes have been defined by BS EN 335-1. These are summarised in *Table 3.3*.

3.8 Extreme hazards

Separating two forms of extreme hazard from the main risks of fungi and beetles has permitted the simplification of *Table 3.3*. For parts of structures that are permanently immersed in saline water (hazard class 5), there are marine boring organisms, such as teredo, that may cause serious damage. In such conditions, timbers with the 'very durable' natural heartwood classification are highly advisable.

For such demanding civil engineering applications, readily available dense hardwoods include Ekki and Greenheart, both of which have properties designated in BS 5268. Occasional infestations by termites have occurred sporadically in parts of the United Kingdom, but these have been in mild climatic locations, where the introduction of the pest is suspected to have been through contaminated import consignments. On the other hand, fungi, to which the table mainly refers, may occur anywhere where the exposure in service and detailing allows the moisture content to exceed

Figure 3.4 Examples of extreme hazard conditions: A: Barmouth Viaduct: this carries the Cambrian Line, a historic railway; following extensive repairs in the 1980s, locomotive-hauled trains again regularly cross; the work included replacing a number of the trestle piles with greenheart timbers because in the old piles, severe attack by teredo – marine borers – had been detected; at the left of the photo is the steel swing bridge, constructed in 1901; the remaining 900 m is the all-timber viaduct, essentially as it was when opened in 1867.

B: a prefabricated modular timber bridge in Cameroon, West Africa, accessing the Korup Forest Reserve. Termites are extremely active here, so the bridge was built using Dahoma, a locally available, naturally durable timber; as an extra precaution, which would be unnecessary in the United Kingdom, after cutting and shaping, all the members were steeped in tanks of hot creosote
Photo A: © Lee Scott
Photo B: © CJM

Table 3.3 Hazard classes in relation to fungal and insect agents, based on BS EN 335-1*

Hazard class	General service situation	Description of exposure to wetting
1	Above ground, covered, dry	None
2	Above ground, covered, risk of wetting	Occasionally
3	Above ground, not covered	Frequently
4	In contact with ground, or fresh water	Permanently
5	In salt water	Permanently

Note
* The table in the standard also addresses marine borers and termites, but here it is simplified – see Section 3.8.

around 20% for a significant time. Also, beetle attack of wood is possible at all moisture contents, but it is only a serious risk to the heartwood where moisture content remains high.

3.9 Hazard classifications for normal bridge components

Most bridge components are likely to fall somewhere within hazard classes 2–4, inclusive. In order to give specifiers more detailed guidance, suppliers of timber protection systems have introduced the term 'use classes', often sub-dividing class 3, as:

3a = Above ground, not covered but coated, e.g. with water-repellent pigmented finishing system

3b = Above ground, neither covered nor coated

With this refinement, and leaving aside the special provisions necessary for immersed marine timbers and termite risks, the 'use' classifications for bridge timbers in the United Kingdom appear as indicated in *Table 3.4*.

3.10 Risk evaluation and decisions

The topic of durability and potential treatments is completed by reference to several European Standards (published in the United Kingdom by BSI). The natural durability of timber species, classified according to the resistance of the heartwood, is indicated by BS EN 350-2. Each species is classified according to its resistance to fungi, insects and the extreme hazards mentioned above. This standard also contains a four-class system of treatability, which is necessary because vacuum-pressure processes are only able to drive the preservative into the heartwood according to its permeability (most species have permeable sapwood). Very broadly, permeability is a function of the density of the timber, although as with all

Table 3.4 Use classes for bridge elements and components

Use class	Situation and exposure	Examples
1	Above ground, covered, permanently dry	Truss or arch elements that are completely encased in ventilated cladding on all four sides of their cross-section. Members within glazed bridge superstructures
2	Above ground, covered, occasional risk of wetting	Many of the structural elements and components that should be protected by well-designed and maintained (open-sided) roofs or with local covers that afford a similar degree of shelter
3a	Above ground, not covered but coated, frequent risk of wetting	Vertical elements and steeply-canted members that allow rapid drainage of precipitation.* Sealed decks with adequate membranes and separate pavement surfaces, e.g. asphalt systems
3b	Above ground, not covered, non-coated, frequent risk of wetting	Cladding comprising sap-free naturally durable species, cladding of preservative-treated timber (= non-surface-coated) or special products, e.g. heat-treated cladding
4	In contact with ground or fresh water, permanent risk of wetting	Exposed decks of small, simple footbridges where the timber boarding, laid with gaps, is easily replaceable. Timbers just above pier caps, abutments and foundations that should nevertheless be detailed with care

Note
* For most bridge elements and components, it is strongly recommended to consider employing coating systems (as distinct from, and possibly in addition to, in-depth pressure preservation). These delay/deter the development of disfiguring surface moulds, algae, lichens and fissures that with time become the seat of more serious degradation.

aspects of this natural material, there are exceptions and variations. However, it is certainly true that most wood species that are classed as naturally 'very durable' are also extremely dense, often equal to or more than 1000 kg/m, and consequently completely impenetrable with preservative. On the other hand, such timbers do not require artificial preservation and are supplied with most if not all of the sapwood excluded.

A guide to durability giving the links between hazard classes and natural durability is contained in BS EN 460, and this leads to the final decision as to whether or not preservative treatment is required. BS EN 599-1 is a performance standard that prescribes the minimum testing requirements and data interpretation for preservatives, while penetrations and retention are defined in BS EN 351-1. Although the European performance standards system in totality seems rather complex, manufacturers' organisations have agreed quite straightforward, quality-assured solutions for all of the common situations shown in *Table 3.4*.

For timber bridge design, *Figure 3.6* summarises the normal risk evaluation and decision-making process.

3.11 Preservative treatment types

There are three main groups of chemical wood preservative systems: waterborne types; light organic solvent-based formulations; and tar oils. In recent years, major advances have been aimed at strict process control,

Figure 3.5 Examples of the 'use classes' for bridge elements and components, as indicated in *Table 3.4*. A: Vihantasalmi bridge. B: Andelfingen bridge. C: Pont de Merle. D: Passerelle Pervou.
Photo A: © CJM
Photo B: © CJM
Photo C: © Wikimedia Commons
Photo D: © CJM

Figure 3.6 Summary of the normal risk
evaluation and decision-making process

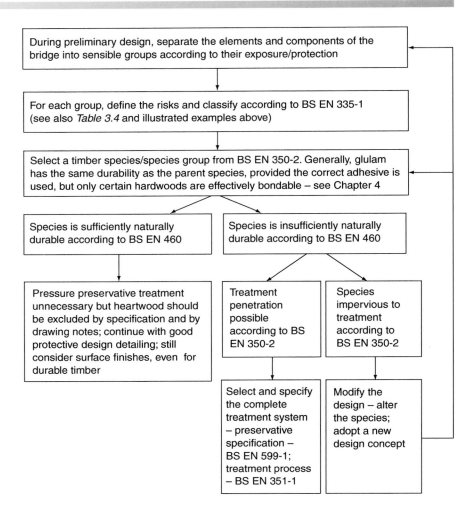

During preliminary design, separate the elements and components of the bridge into sensible groups according to their exposure/protection

For each group, define the risks and classify according to BS EN 335-1 (see also *Table 3.4* and illustrated examples above)

Select a timber species/species group from BS EN 350-2. Generally, glulam has the same durability as the parent species, provided the correct adhesive is used, but only certain hardwoods are effectively bondable – see Chapter 4

Species is sufficiently naturally durable according to BS EN 460

Species is insufficiently naturally durable according to BS EN 460

Pressure preservative treatment unnecessary but heartwood should be excluded by specification and by drawing notes; continue with good protective design detailing; still consider surface finishes, even for durable timber

Treatment penetration possible according to BS EN 350-2

Species impervious to treatment according to BS EN 350-2

Select and specify the complete treatment system – preservative specification – BS EN 599-1; treatment process – BS EN 351-1

Modify the design – alter the species; adopt a new design concept

avoiding risks to the health of treatment personnel, installers and the general public, and ensuring quality-controlled efficacy during the service life of the timber. New treatments have been developed that offer both water-based and organic solvent-based methods. The latter are generally more commonly used as treatments for joinery and similar items. In general terms, all of these types contain blends of active compounds that function both as fungicides and insecticides. A fixing agent is also included, so that while drying occurs (often nowadays itself a carefully timed, computer-controlled process), alterations take place in the chemical consistency of the formulation, binding it to the wood and making the timber resistant to leaching.

For 'use' categories 3, 4 and 5 in structures, either high-pressure waterborne formulations or tar oils are generally specified. In January 2003, an EEC directive was issued placing restrictions on the use of the older chromated copper-arsenate (CCA) systems that were the mainstay of this application. Although an EEC Derogation (exception) was included in the directive permitting continuation for industrial installations, including bridges (and items such as avalanche barriers), the treatment companies have now all introduced satisfactory alternatives. One large producer bases treatments on aqueous solutions of copper salts with ammonia or amines (broadly described as ACQ products). Another offers impregnation with formulations of copper and triazole biocides, optionally adding water repellents and in-depth colour pigments.

Figure 3.7 A newly installed pressure treatment plant undergoing trials
Photo © Osmose

3.11.1 Processes

The actual treatment processes vary in detail, but broadly they comprise combinations of loading a charge of timbers into a cylinder, pumping a vacuum, filling the vessel with preservative solution that is forced into the timber under pressure, then drawing off the excess liquid. The timber then goes into a separate zone where it is dried, during which the moisture content is regulated and fixation checked. There is a high degree of automation, and bunding of the plant to prevent seepage into ground water tables. Similar comments and procedures apply to the processes that inject creosote. This is a very old method, but it is still valid for heavy-duty applications in large bridge elements that cannot easily be touched by the general public who use the bridge when it is put into service (see Flisa, Tynset and Evenstad Bridges in Norway, for example). Because of their very demanding design working life classifications, the timbers in these large bridges have been double-treated, as described in the case studies. The CCA process that was initially applied to the individual laminations before they were bonded-up into glulam can be replaced with the types of treatment described above, e.g. ACQ methods.

3.11.2 Surface treatments

Surface treatments are available, based on broadly similar formulations, but compounded to give a greater degree of adherence, e.g. as pastes or emulsions. Generally these are insufficient in their own right for anything beyond use class 2, but they have a place where a moderate amount of cutting, drilling or notching is unavoidable after delivery; also for repair work when the ends of parent timbers can be treated before splicing-on replacement parts.

3.11.3 Metal-free preservatives

Some types of metal-free wood preservatives have for some time received independent approvals and have achieved significant sales levels. Notably, this includes boron compounds. The attraction is the reduction in difficulties at the disposal or recycling stage, because of extremely low levels of mammalian toxicity. From the perspective of use in applications such as bridging, the snag is that such chemicals are waterborne, and thus easily leached out by repeated wetting. Also, for bridges over streams where fish-life is conserved, boron solutions have a harmful effect on the small crustaceans upon which they feed. Research continues, with the aim of finding affordable additives to improve boron fixation, and also to develop completely natural protection systems. The latter could include plant-derived, or tree leaf or bark compounds.

3.12 Relating the classifications

As we shall see in Chapter 7, structural design requires the engineer to determine the service class for a component, taking account of where it is situated, hence as a function of its location within the complete structure – in this case the bridge. As has been shown above, classifications are also made of the degrees of protection afforded to the various parts of a timber bridge, and again these may not be identical for all of its items. Finally, hazard categories and or use categories have been discussed. While it is not possible categorically to state that all three of these classifications intermesh with perfect precision, there is clearly a resemblance between the high-moisture ambient environments, the more exposed components and the high-hazard situations. Thus the summary in *Table 3.5* suggests where the designer is likely to find that these, and the lower risk, drier situations, tend to relate to one another.

For use classes 3a, 3b and 4, the designer will often look to gain additional durability from species in durability classes 2 and 3 by means of pressure preservative treatment, as discussed above. The amenability of the proposed species (preservative penetration and retention) will then need to be examined. Sections of BS EN 350-2 deal with treatment classes. If considering a species that is rated as difficult or extremely difficult to treat then suppliers of pressure preservative systems should be consulted. Because of the necessity to treat all timber structures in parts of the world where termites and tropical insects are a risk, methods have been developed to treat even quite resistant timbers, but these may not easily be sourced in the United Kingdom.

3.13 Construction protection arrangements

These are the first line of defence in ensuring durability. Right at the start of the initial design, the following guidelines should be observed. As already shown, a constant, high moisture content in the structure is likely to lead to fungal attack and subsequent problems including insect pests. The principal objective therefore is to eliminate this possibility. To ensure durability by maintaining a dry timber bridge:

- The minimum quantity of water should fall upon the structure.
- Water that does fall must drain away as rapidly as possible.
- Liquid water and dirt must not become entrapped in any of the details.
- All parts of the structure should be well ventilated.

Table 3.5 Summary, showing the general relationship between structural service classes, use classes, and choices for natural durability/ preservation treatment

Structural service class*	Use class†	Natural durability class‡ In each indicated category, a timber species from any of the durability classes shown below may be selected	Treatment§
1 or 2	1 Covered, permanently dry	1 to 5 inclusive§§	Only surface treatments – optional
2	2 Covered, occasionally wetted	1 to 4 inclusive	Surface treatments – optional; specify 'sapwood excluded'
3	3a Coated, frequently wetted	1 to 3 inclusive	Surface treatment required, by definition; to provide UV protection, fungicide and decorative finish
3	3b Uncoated, frequently wetted	1 and 2 (and 3 for cladding) – or pressure treat	Cladding from sap-free timber, durability class 1–3; structure as below
3	4 In/near ground or freshwater contact	1 and 2 – or pressure treat	Essential, unless using Class 1 or 2 durability, in which case pressure treatment unlikely to be possible

Notes
* Service classes defined in BS EN 1995-1.
† Use classes = hazard classes, defined in BS EN 335-1.
‡ Natural durability classes defined in BS EN 350-2.
§ Treatment classes defined in BS EN 350-2.
§§ Note the numbering system for 'natural durability against fungi' – Class 1 = 'very durable', Class 5 = 'not durable'.

Once these points have been observed, it is good discipline in preliminary design to re-visit the risk evaluation explained above, and to reconsider the options for timber species, treatments and finishes. Only then should concept design continue (Chapter 5), followed by full structural design (Chapters 6 and 7). The following sections examine and illustrate specific aspects of construction protection.

3.14 Covered bridges

Several of these have already been illustrated, and in Chapter 2 we saw how a roof can ensure that a timber bridge lasts for centuries. This therefore may well be considered a worthwhile investment, even though it obviously creates a more expensive structure than, for example, open trusses or arches. To ensure that the principal elements are adequately sheltered there are rules for roof overhangs and pitches that must be observed. These are discussed in Chapter 6.

For a site where normal precipitation occurs, rather than a very exposed coastal site, the key elements can be designed as service class 2 members (Chapter 7), provided that an angle of shelter of at least 30° is created. For exceptionally exposed sites the further precaution may be to utilise a 'durable' or 'very durable' timber species, but still to consider the service class 2 situation for the calculations. Transoms and stringers that are fully sheltered under a sealed deck may also be considered to be 'roofed' and therefore to qualify for this design assumption. *Figure 3.6B* shows an example.

3.15 Cladding and local covers

Another important option, cladding and local covers, may either comprise almost complete timber covers (*Figure 3.8*) or, as also shown in this illustration, it may take the form of louvres with more open gaps between boards. In an unroofed bridge, unless the selected species is from the 'very durable' category of natural resistance, horizontal upper surfaces such as the top edges of beams, girders and arches should *always* be protected either with cladding or metal.

3.15.1 Arrangements of board cladding

Cladding with timber boards or 'modified wood' (see Chapter 4) will generally be of 19–28 mm thickness (depending on the profile) and

Figure 3.8 Examples of cladding.
A: principal members completely encased in ventilated timber cladding for kingpost trusses in a major road structure – Vihantasalmi Bridge.
B: larch louvre screening on a modern covered trussed-girder road bridge – Andelfingen, Switzerland
Photos © CJM

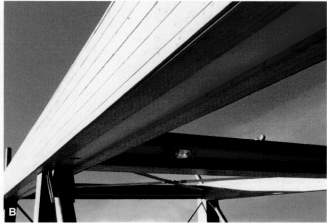

Figure 3.9 A: Horizontal feather-edged larch cladding and metal covers over the arches, to protect this bridge near Aachen, manufactured by Schmees & Lühn.
B: Louvred cladding on the glulams of the under-tied beam bridge 'Pont de Bouix', Commune de Barnas.
Photo A: © A. Lawrence
Photo B: © J. Anglade

75–150 mm width. The moisture content of the boards alters through seasonal fluctuations and exposure, so this can lead to undesirable distortion unless the profiles, joints between boards, supports and fixings are well designed and executed. Ideally, quarter-sawn boards are the most stable in terms of avoiding cupping, but it is rarely feasible, either in terms of cost or availability, to insist on this method of conversion. Designs that allow for moisture content change and consequent timber movements are given in *TRADA External Timber Cladding*. Horizontal and vertical board arrangements are covered (*Figure 3.9*), both useful in bridge design. Although principally aimed at cladding for the protection of buildings, most of the principles are similar for bridges.

To shelter major structural members, vertical or inclined board cladding is usually chosen – see, for example, *Figure 3.9A*. This achieves the rapid surface water run-off that is desired. However, the alternative type of system shown in *Figure 3.9B* in this illustration is also effective, acting to filter driving precipitation in a similar way to traditional Yorkshire boarding on farm buildings.

To attach cladding to its supports, and the latter to the main structure, austenitic stainless steel fasteners are recommended, and it is well worth considering investing in the slightly higher cost of screws rather than nails, since these are likely to facilitate maintenance at a later stage. Corrosion-protected clip fixings are another possibility, and suppliers of timber engineering hardware provide a wide range.

Support battens for cladding that are applied over flat structural surfaces should be arranged so there are paths for air circulation and condensation drainage behind the boards (*Figure 3.10*).

3.15.2 Louvred cladding

As protective detailing for many bridge elements, louvred cladding is an attractive and effective option. Most of the principles outlined above are equally applicable. Shading deck edges and aesthetically significant principal members with louvres brings advantages of reducing fissuring through UV-light exposure as well as filtering precipitation and encouraging run-off.

The support arrangements for the louvres need to be relatively strong and robust, since they normally need to be located at wider centres than normal cladding battens. Also, the louvres themselves are acted upon by wind and self-weight.

Key design points for louvres include the following:

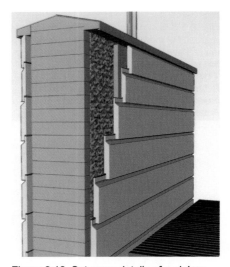

Figure 3.10 Cut-away details of a glulam 'trough bridge' – the main beams are clad with shiplap boards, mounted on battens, behind which is breather membrane to keep the glulam side faces clean. The voids between the battens and below the sill (with drip) are open at the bottom
Drawing © CJM/TRADA Technology

• Specify good-quality material of a species that is at least 'moderately durable' (Class 3 BS EN 350-2) – alternatively, a proprietary form of 'modified wood' may be suitable – but check for independent approval certificates!
• The strength and stiffness of the louvre members should be considered, even though they are not load-bearing – very light timbers such as Western red cedar may need special support arrangements.
• Water staining and stress corrosion should be avoided by matching the species to the fastener metallurgy and avoiding chemical and electrolytic interaction.

3.15.3 Metal covers

Metal covers formed from stainless steel and various alloys of aluminium, zinc and copper are an important tool in the designer's armoury.

As indicated above in relation to cladding, metal fixings, brackets, clips, etc. should be selected bearing in mind their chemical and electrolytic compatibility both with the wood and with one another. Thin metalwork that completely covers a flat timber surface should not make full contact, but is usually better lifted off the structure. This can be achieved by blocking up thin support boards of a material similar to the type used for cladding, and then attaching the metal strips to it.

Where an open-boarded deck type is used, the water is deliberately encouraged to drain through, by means of gaps between each board and gaps at the edges. Thus the members beneath, unless of a very durable species, must all be provided with suitable protection (*Figure 3.13*). Further details of this type are shown in Chapter 6. Such covers installed locally, for example at the feet of struts and columns, are often detailed to contain water channels, in a similar manner to the design of flashings in building junctions. A good example can be seen in *Figure 3.8A*, showing the protection over the base supports for the A-frames of Vihantasalmi Bridge.

3.15.4 End grain protection

The best solutions for end grain protection generally involve metal covers – see, for example, *Figure 3.14*. Attaching another moulding of timber whose grain runs across the width of the main member can cover smaller end grain items that would otherwise be exposed. It is important to ensure that ventilation is included, so that by adding such items the situation is not worsened by creating traps for wet dirt. To protect valuable furniture and joinery timbers when these are held in stock, formulations of epoxy resin are often bonded to the end grains, but at present there is not convincing evidence that this method of protection can be used permanently on exterior structures.

3.15.5 Parapets as protection

Depending on the arrangement of the bridge, it may be possible to incorporate the parapet and handrail into the protection plan (*Figure 3.15*).

Figure 3.11 Close-up of the louvred cladding on Andelfingen Bridge. Note the excellent quality of the material; also the neat counter-sinking of the parapet rail fixings. This ensures not only good appearance but also durability
Photo © CJM

Figure 3.12 Examples of metal covers for protection. A: Swiss open bridge with glulam arches and through deck.
B: Norwegian bridge for pedestrians and light traffic – the top chords of the Warren girders are covered; the inserted steel plate connections in the foreground end-strut have open-bottomed drainage slots. The protection scheme for the main timbers is double pressure-treated European redwood (pine); note also the mesh protection against objects falling onto the highway below
Photo A: © CJM
Photo B: © TRADA Technology

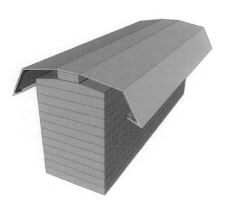

Figure 3.13 A typical metal cover of the type used over a glulam longeron, below an open-boarded deck
Drawing © CJM/TRADA Technology

Figure 3.14 End grain protection. A: Pont de Crest, metal covers over the raking parapet struts ensure that precipitation runs down and drops off the bottom without nearing the end grain of the struts or lower joists.
B: Thalkirchener Bridge, Munich also includes protection in a similar zone, but here the upper edges of the joists are covered, while their outer ends are cut at a steep inward-raking angle, again to ensure water shedding
Photo A: © CJM
Photo B: © STEP

Figure 3.15 A: the handrail is sloped laterally towards the outside of the parapet. An angle of 5% or more is recommended. The illustration also raises the question of where protection should stop! But with little further expense, another board could be added, with its face across the ends of the cladding boards.
B: details of the well-known and spectacular tension-ribbon bridge at Essing, now 20 years old, where the parapet post can be seen to be 'stood off' from the beams, using round steel spacers as recommended; also a protective cladding has been added after recent inspections to provide further protection to the vital tie beams
Photo A: © CJM
Photo B: © A. Lawrence

This is especially likely if through-beams or girders are involved. Ventilation should be provided under the top handrail board, and also in the joints and edges of the cladding. 'Best practice' details for such aspects are included in *External Timber Cladding* by Patrick Hislop, published by TRADA Technology.

To ensure that maintenance attention is not too regular, and use by the public does not become unpleasant, handrails require protective detailing in their own right. This aspect is taken further in Chapter 6.

3.16 Connections

For connections, effective construction protection arrangements are vital (*Figure 3.16*). Badly designed bridges that have deteriorated too soon are often found to have failed principally in this area – see, for example, *Figure 8.5*. Contact points between ground level or supports and the superstructure are especially vulnerable, and should always be raised, drained and ventilated, and possibly covered. Because good timber design should take into consideration the arrangement and nature of the connections at an early stage, so also protection needs to be thought of right from the start. The interrelationship between structural performance and detailing is such that these matters are explored further in Chapter 7.

3.17 Junctions

Several instances where junctions occur between one timber component and another have already been described – in the parapet and handrail zones, for example. The aim is to minimise contact faces. Between decking planks and beams or stringers, there are a number of improved ways of making the connections. Illustrations are included in Chapter 6. Junctions with other materials also occur at positions other than the structural connections – for example, where cladding is taken down to surrounding concrete. These are generally formed with pitched and tucked-in metal flashing, in a similar manner to such details in building structures.

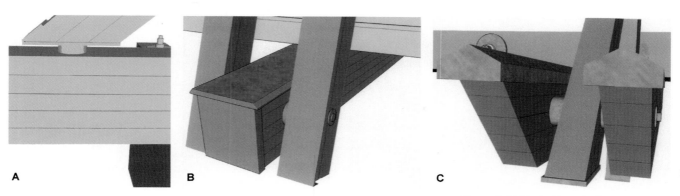

A   B   C

3.17.1 Deck-to-approach junctions

At the transition zones, where the traffic and users pass onto and off the bridge, vulnerable junctions are inevitable. A variety of proprietary bridge seals can be adapted for use with timber bridges, and these provide a convenient means of achieving the transition from the timber structure to the approach pavements.

Construction protection is vital, and all of the points discussed above regarding natural durability, pressure preservative treatment and partial covering should be considered. All of the timber elements must be a sufficient distance from the concrete or earth to avoid wetting and water trapping. The end grain regions of members are, of course, particularly vulnerable. The concrete supports should be designed so that run-off is achieved, with the collected water drained away, without risk of soaking the timber. All timber elements must be well ventilated and protected from potential wetting from above. Supports should be detailed so that silt and debris cannot accumulate. Access must be given for cleaning channels, gutters and areas below gratings.

3.18 Decks

Construction protection design for decks begins by considering whether the structure is to permit water to percolate through, or whether it is to be sealed. Poor designs often result from falling between two stools and doing neither successfully! Essentially, the alternatives are:

*   Well-ventilated and drained open timber decking, generally using open-gapped boards for which the risk evaluation and decision-making process summarised in *Figure 3.6* is applicable.
*   Timber or timber composite decking that is protected and sealed using several layers. Generally these entail a membrane above the timber plate, an under-layer of waterproof material such as a bitumen composition and a wearing surface for the traffic.
*   Use of reinforced concrete in conjunction with a timber deck plate, or above timber stringers and transoms.
*   Sealing and shear-keying to the timber, topping with screeds, etc. in accordance with normal concrete technology.

Deck design is considered in more detail in Chapter 6.

Figure 3.16 Typical 'stood-off' connections at member junctions, to avoid water trapping, in conjunction with local metal covers.
A: the bent-over bracket at the top of the parapet strut;
B: the pair of diagonal struts rising each side of a single transom;
C: the pair of transoms with swept-back cut ends and a single diagonal strut between. The spacers are formed with short lengths of galvanised round hollow steel containing bolts and timber connectors
Drawing © CJM/TRADA Technology

Figure 3.17 An example of a well-detailed and protected deck-to-pavement junction for a pedestrian and cycle bridge.
Key: (1) membrane and two-layer bituminous aggregate pavement;
(2) reinforced concrete abutment, note slopes for drainage – internal details not shown; (3) proprietary bridge joint with elastomeric seals and membranes;
(4) castellated durable hardwood decking;
(5) corrosion-resistant metal cover over glulam – extends to beam ends; (6) glulam main beams, protected; (7) transverse steel I-beam, protected and anchored, with height-adjustable arrangement for timber-bearing plates; (8) corrosion-resistant metal flashing over end grain of glulam
Drawing © CJM/TRADA Technology

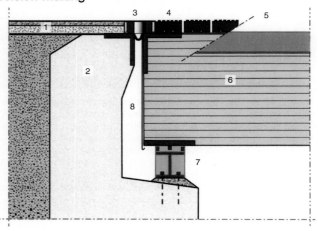

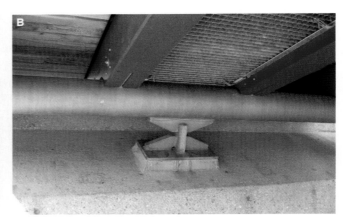

Figure 3.18 A covered pedestrian and cycle bridge at Traunreut, illustrating several of the principles given in *Figure 3.17* – including the bank seat caps that are well sloped for water shedding and the easily accessible grille for drainage and debris collection. Note also the water-repellent stain finish on the glulam, some of which is of round-turned section. This is a nonsealed deck – water can filter freely through the slightly spaced deck boards, so the longerons (red, in A) have metal covers as in (6) in *Figure 3.17*
Photos © Ingenieurbüro Miebach

3.19 Constructional protection example: Pont de Merle

Five multiply strutted frames provide the supports for the 57.7 m long high-level decked road bridge shown in *Figure 3.19*. An unusual feature is that the principal compression elements consist of a built-up T-section made from glued laminated French-grown Douglas fir. These slender members vary in length between 12 and 25 metres. Architecturally, the frames were conceived as zig-zagged lighting strokes.

The deck comprises five sets of continuous glulam beams, each comprising a pair of timbers that are rigidly braced against one another. At positions outboard of the central 'X-struts', continuity nodes are formed using true pins, so that in effect this is a cantilevered arrangement, with a central span and further moment-carrying flexural members running from the upper abutments to these two-thirds length pin positions. The timber deck structure is canted laterally at a 2% slope for drainage, and on top is a modular reinforced

Figure 3.19 A: The Tours de Merle are a group of mediaeval castles in the Corrèze Département of France, situated in a beautiful gorge of the River Maronne. They are a classified historic monument, and used to be approached by an impressive suspension bridge, built in 1852. This carried all the road traffic, but after having a weight limit of 12 tonnes for some years, it was eventually declared completely unsafe and dismantled. In view of the excellent French-grown Douglas fir that is available locally, the architect responsible for the new 'Pont de Merle' elected to use timber, complementing the protected landscape.
B: The result is a magnificent new timber road bridge with unlimited load capacity that soars spectacularly 30 m over the riverbed, on this steeply banked and heavily wooded site. Harnessing the skills of local carpenters, boldly angled struts and near-vertical columns echo the precipitous limestone cliffs of the region. Special design considerations addressed ease of manufacture and minimisation of transportation costs, as well as protective design measures to ensure full durability and maximum longevity with ease of maintenance
Photo A: © Michotey JL
Photo B: © Wikimedia Commons

concrete diaphragm. To connect this to the sub-structure, serrated steel shear pins are bonded into the glulam members using an epoxy resin adhesive system. A layer of moisture-resistant neoprene is included, and the traffic running surface is bitumen-based. The Douglas fir glulam is sap-free, class GL30, while the pedestrian walking surface is of azobé (known in the UK timber trade as ekki) – a very durable and hard-wearing tropical timber. For the handrails, another 'very durable' (EN 350-2 classification) tropical hardwood – iroko – is used.

3.20 Principal timber species in relation to durability design

Table 3.6 shows a summary of the main timber options for each use class and situation in the bridge. It should be noted that this is not intended to be comprehensive, but is to assist designers in preliminary choices of materials that are readily available in the United Kingdom. A large number of alternative species may be available, even in the United Kingdom, while elsewhere there is considerable further scope. In general, the choices should be examined using the philosophy and references provided here and in Chapter 4, together with the case studies illustrated in Chapter 9.

Figure 3.20 The struts and column bases are articulated as true pins. Because of the steep terrain, the four base supports are at varying levels. In all cases the reinforced concrete foundation caps are canted inwards significantly, to resist the outward component of the reaction thrusts, and to form part of the protective design solution, ensuring rapid drainage of precipitation away from the timber bases. As can be seen from this detail, the upper elements of the 'T-sections' help to protect the regions where inserted plates run into the end grain. Also, the whole structure is protected with a water-repellent and fungicidal pigmented stain finish. At the upper levels, the sealed deck offers considerable protection to the main carriage beams, as is always the case with correctly detailed high-decked strut or arch bridges

Table 3.6 Summary of the main timber options for each use class*

Use class	Situation in bridge	Timber species/species group	Treatment
1	Fully covered, permanently dry	Spruce or any other *sp.* with equal or better natural durability, including all in the categories below	Surface treatment – pigmented finishes providing UV-light protection + fungicide against algae, lichens and moulds
2	Covered structural components with likelihood of occasional wetting	Caribbean pine; Douglas fir; European redwood; larch; Scots pine; American white oak; European oak; sweet chestnut All of those tropical hardwoods listed in *Table 4.3*	No pressure treatment essential for this use class provided sapwood is excluded by specification; surface treatment advisable but not essential
3a	Cladding – coated but likelihood of frequent wetting	Stable, and probably lower-density species selected from above list, plus Western red cedar 'Modified wood' products with independent approvals for use as cladding	Surface treatment essential to qualify for this use class
3a	Structural components – coated but likelihood of frequent wetting	Caribbean pine; Douglas fir; European redwood; larch; Scots pine; American white oak; European oak; sweet chestnut All of those tropical hardwoods listed in *Table 4.3*	Surface treatment essential to qualify for this use class Pressure preservative treatment optional – check feasibility
3b	Cladding – non-coated; likelihood of frequent wetting	Caribbean pine; Douglas fir; larch; Western red cedar; American white oak; European oak; sweet chestnut All of those tropical hardwoods listed in *Table 4.3*	Only use 'moderately durable', 'durable' or 'very durable' species with sap excluded by specification
3b	Structural components – non-coated; likelihood of frequent wetting	Caribbean pine; Douglas fir; European redwood; larch; Scots pine; American white oak; European oak; sweet chestnut All of those tropical hardwoods listed in *Table 4.3*	Pressure preservative treatment or use of 'durable'/'very durable' species essential

Note

*This is not intended to be comprehensive, but is to assist designers in preliminary choices of materials that are readily available in the United Kingdom. It should be read in conjunction with Chapters 3 and 4.

4 Materials

4.1 Introduction: the generic range

This chapter presents the materials that are used in the design and construction of modern timber bridges. The correct choice of timber should be considered carefully at an early stage. If necessary, this should be undertaken in consultation with an architect or engineer who specialises in the subject. Potential manufacturers should also be consulted promptly, since it is important to realise that unlike the situation with bridges of concrete and steel, there is not a large supply industry, particularly within the United Kingdom. In Chapter 3 we considered the implications of the natural durability of timber, its potential for various forms of preservative treatment and additional means of ensuring longevity through protective design. Here the natural durability classifications of commonly used timber bridge species are tabulated, alongside their treatability classes and other definitions and properties.

Materials formed from wood (*Figure 4.1*) are broadly categorised as follows:

- solid timber (optionally machined, after sawing, on two or four lateral surfaces)
- glued laminated timber (glulam) – made under factory process control from selected grades of solid timber, with finger joints in the laminations that are also strictly controlled
- mechanically laminated timber – factory- or workshop-made from solid timber
- laminated veneer lumber (LVL) – strictly controlled manufacture, rotary peeled veneers.

Figure 4.1 Examples of the generic range of materials – within this broad framework, numerous options are available.
A: solid timber – in this case sawn British-grown Douglas fir;
B: glulam with proof-loaded finger joints;
C: mechanically laminated greenheart;
D: pressure preservative treated STC T-beams
Photos A–D: © CJM

Glulam and LVL can be described generically as structural timber composites (STCs). In addition to LVL, there are further types of STC, but the others are less frequently used for bridging. Certain forms of structural wood-based panels may also be used, especially plywood, provided this is of a species and lay-up having sufficient durability – achieved either naturally or through adequate preservative treatment processes. Finally, some innovative timber engineering materials such as structural multi-layer board are beginning to be employed.

All of these families of materials need to have attested, quality-assured properties so that, for structural design, the engineer can make reference to them via the standards developed through the European Committee for Standardization – CEN – i.e. those published in the United Kingdom as BS ENs. For whichever type is intended, it is important to check on the conformity at an early stage in design, and then to ensure that the choice is obeyed during manufacture and installation. Some of the families of wood-based materials used generally in the construction industry, particularly for short-term applications such as formwork, may not have the necessary durability classification that would make them suitable for reference via a Harmonized European Standard for use in bridges.

4.1.1 Steelwork
It is usually necessary also to include steel in one form or another. This may either be one of the plain carbon structural grades, or a suitable stainless steel type. Instances where steel is used in timber bridge design include:

1 fasteners and connectors for the timberwork
2 plates and other fixings at connection nodes in the structures
3 pre-stressing tendons, e.g. for laminated decks.

In bridges, resistance to corrosion is particularly important, and this is addressed by codes such as BS EN 1995, BS EN 1993 (steel) and BS EN 1992 (concrete – for pre-stressing bars and cables).

Sometimes, actual structural members in steel may be combined with timber elements, resulting in bridges that are effectively of mixed construction (*Figure 4.2*). This can make good sense – for example, where there are high-tensile forces to be carried. The fact that both the timber and the steel Eurocodes have their basis of design in BS EN 1990 now makes it easier and less error-prone for engineers to investigate this approach, whereas previously two different safety systems existed in the respective design codes.

As we shall see below, bridge frames may also be constructed using carpentry techniques, with cut joinery for the connections, and these may avoid the use of metals altogether. Where this decision is appropriate, generally in smaller forms of bridge, this may be considered an advantage.

Figure 4.2 Pont du Bouix, a bridge for light vehicular traffic in the Ardèche – structural steel ties in a construction that is principally timber (and weather-protected using sweet chestnut louvres). By taking out the horizontal thrust, this structural arrangement avoided an expensive replacement of the existing piers that previously carried a stone arch
Photo © J. Anglade

4.2 The main options

Even if the option of adding pressure preservative treatment is adopted, the species of timber selected should always have a degree of natural durability – while protective design measures are very important, they are unlikely on their own to transform a bridge made from an unsuitable timber. The fundamentals of achieving durability have been discussed in Chapter 3, where the hazard definitions have also been examined. Natural durability classifications are indicated in this chapter, so that they can be tabulated together with the mechanical properties.

Further important issues to consider are the available cross-sectional sizes, lengths, amenability to treatment if required and, in the case of STCs, economy in manufacturing and perhaps size limitations for treating the components of the bridge. For quite short spans (a quantitative indication of which is provided in Chapter 5), either solid sawn softwoods or hardwoods are generally an option. Broadly speaking, larger sections and lengths are available in hardwoods than in softwoods, and choices of both temperate and tropical broad-leaved timbers are available.

For more than 60 years, the main alternative to solid timber has been glulam. This is manufactured both from softwoods, with which it is possible to specify species that are in the 'moderately durable' category, and also from hardwoods, some of which can be rated as 'durable'. For these, however, it is necessary to be very sure of the precise species, and to ascertain that it is amenable to bonding with the appropriate adhesives – there are certain well-established species that fulfil the requirements. Quite recently, other forms of STC, notably LVL, have been introduced. These are generally manufactured from wood species such as spruce and silver fir, which have lower natural durability, but treatments have been developed that involve pressure preservation and excellent penetration into the material. This is a specialised process, for which manufacturers' advice should always be sought.

4.3 Solid timber

The type and quality of solid timber that is available varies from small sections of softwood or hardwood for simple short-span beam bridges, through larger sections in temperate or tropical hardwood, to quite long lengths (e.g. up to around 12 m) in specialist tropical hardwoods. These are required for the biggest members, such as masts for cable-stayed bridges (*Figure 4.3*).

Elsewhere in the structure, species commonly employed in the solid form include softwoods such as Scots pine and larch (*Table 4.1*), temperate hardwoods, such as European oak (*Table 4.2*) and tropical hardwoods such as ekki, iroko and opepe (*Table 4.3*).

Although strength classes are sufficient to categorise the required combinations of species and structural grade for design calculations, because of the necessity for natural durability and, in some cases, good appearance it may be insufficient simply to specify the choice of the timber in terms of the strength class. This is further considered below – under *strength classes.*

Figure 4.3 Because of its availability in exceptional lengths and cross-sections, it was possible to use Basralocus (*Table 4.3*) for the masts of the Enschede Bridge in the Netherlands – a canal crossing for pedestrians, cyclists, horse riders and emergency vehicles
Photo © CJM

Table 4.1 Softwood timber species commonly available in the United Kingdom for bridge construction

Standard name/origin	Botanical species	Air dry density kg/m³	Durability class (EN 350-2)	Size class	Treatability class (EN 350-2)	
					Heartwood	Sapwood
Douglas fir – imported (I) and British grown (BG)	Pseudotsuga menziesii	(I) 510–550	Moderate (3)	Fairly large	4	3
		(BG) 470–520	Slight (4)		4	2–3
European and Siberian larch	Larix decidua Larix kaempferi; Larix x eurolepis Larix occidentalis	470–650	Moderate (3) to slight (4)	Normal	4	2
Caribbean pine	Pinus carribea Pinus oocarpa	710–770	Moderate (3)	Fairly large	4	1
European redwood, Scots pine	Pinus sylvestris	500–540	Moderate (3) to slight (4)	Normal	3–4	1
Southern pines group (USA)	Pinus elliottii Pinus palustris Pinus taeda Pinus echinata	650–670	Moderate (3) to slight (4)	Fairly large	3	1

Table 4.2 Temperate hardwood timber species commonly available in the UK for bridge construction

Standard name/origin	Botanical species	Air dry density kg/m³	Durability class (EN 350-2)	Size class	Treatability class (EN 350-2)	
					Heartwood	Sapwood
American white oak	Quercus alba; Q prinus; Q. lyrata; Q. michauxii	670–770	Durable (2) to moderate (3)	Normal construction (2)	4	2
European oak	Quercus robur; Q. petraea	670–760	Durable (2)	Normal construction (2)	4	1
Sweet chestnut	Castanea sativa	540–650	Durable (2)	Normal construction (2)	4	2

4.3.1 Dimensions

Sizes of solid timber vary significantly according to the species and geographic source, so in *Tables 4.1*, *4.2* and *4.3* dimensional ranges are indicated by means of a size classification. Standard lengths of most softwood imports run up to about 7.2 m. Locally sourced British-grown softwoods such as Douglas fir and larch may be available in considerably longer lengths, but this is always subject to specific arrangements. For hardwoods, the whole subject of dimensions – both cross-sections and lengths, is much more dependent upon the type (temperate (e.g. oak) or tropical (e.g. ekki)), as well as the precise source.

Generally for relatively short spans, solid European oak-beamed footbridges (*Figure 4.4*) remain popular with both owners and the public. Sadly, this is probably the only type of timber bridge for which it may be said that the United Kingdom has an ongoing tradition! Carpentry techniques, possibly assisted by a modicum of discretely hidden stainless steel plate-work, can provide pleasing structures of this genre. More details on the bridge shown in *Figure 4.4* are given in the publication *Green Oak in Construction*, where it is included as a case study.

Figure 4.4 A closer view of another European oak bridge fabricated by carpentry, showing the pegged joints – at Polesden Lacey National Trust Garden, in Surrey
Photo © CJM

Table 4.3 Tropical hardwood timber species commonly available in the United Kingdom for bridge construction

Standard name/origin	Botanical species	Air dry density kg/m³	Durability class (EN 350-2)	Size class	Treatability class (EN 350-2)	
					Heartwood	Sapwood
Balau – Malaysia	*Shorea* spp. *Incl. S. glauca; S. laevis; S. Maxwelliana*	700–1150	Durable (2)	Large construction (1)	4	1–2
Basralocus – South America	*Dicorynia guianensis*	720–790	Very durable (1)	Large construction (1)	4	2
Dahoma – West Africa	*Piptadeniastrum africanum*	610–710	Durable (2)	Normal construction (2)		
Ekki (Azobé) – West Africa	*Lophira alata*	950–1100	Very durable (1)	Large construction (1)	4	2
Greenheart – South America	*Ocotea rodiaei*	980–1150	Very durable (1)	Large construction (1)	4	2
Iroko – West Africa	*Milicia excelsa* (formerly *Chlorophora excelsa*); *M. regia*	630–670	Very durable (1)	Normal construction (2)	4	1
Jarrah – Western Australia	*Eucalyptus marginata*	790–900	Very durable (1)	Normal construction (2)	4	1
Kapur – Malaysia	*Dryobalanops* spp.	630–790	Durable (2)	Normal construction (2)	4	1
Kempas – Malaysia	*Koompasia malaccensis*	850–880	Durable (2)	Normal construction (2)	3	1–2
Merbau – Malaysia	*Intsia Bakeri; I. bijuga; I. palembanica*	730–900	Durable (2)	Normal construction (2)	4	
Opepe – West Africa	*Nauclea Diderrichii*	740–780	Very durable (1)	Normal construction (2)	2	1
Purpleheart – South America	*Peltogyne* spp.	830–880	Durable (2)	Normal construction (2)	4	1
Teak – tropical plantations in West Africa and Asia-Pacific Region	*Tectona grandis*	650–750	Very durable (1)	Normal construction (2)	4	3

Regular construction and joinery hardwoods such as iroko, which may be suitable for cladding or decking, are easily available in lengths of up to 6 m. However, certain heavy-duty civil engineering hardwoods (such as greenheart and basralocus) are available in much longer lengths. Subject to advance notice, it may be possible to obtain these in lengths as great as 20 m, and sections of up to 450 or even 600 mm square. This has already been illustrated in *Figure 4.3*. As with the home-grown softwoods, local hardwoods such as European oak and sweet chestnut should always be sourced through a dialogue with experienced sawmillers and their carpenters (*Figure 4.5*).

Figure 4.5 A: inspecting a British-grown oak log immediately after felling to determine whether carpentry-quality material can be obtained.
B: a stack of sweet chestnut logs in the Kentish Weald – a valuable but under-utilised resource that may be converted into durable glulam
Photo A: © Forest Life Picture Library
Photo B: © CJM

4.3.2 Conformity

For structural design with softwood or hardwood in solid timber form, the essential Harmonised Standard for compliance with BS EN 1995-1-1 is BS EN 14081-1. In essence, the standards treat timber as a naturally variable material, which it is necessary to strength-grade according to formal quality-assured procedures. Hence test samples of solid structural timber (softwood or hardwood) are selected and the population that is to be sampled is expected to be capable of being identified at all stages of production and supply, including the time when it reaches the factory or construction site.

The actual basis for visual strength grading is stated in BS EN 14081-1, since it is a requirement that it must be carried out, but the precise procedures and defect-measurement methods are still provided by national standards. For instance, the United Kingdom and many of its suppliers in the Nordic region operate visual strength-grading for softwoods against BS 4978. BS 14081 Parts 2–4 cover machine strength-grading in great detail, and these aspects now apply uniformly throughout the CEN zone without subsidiary reference to any national alternatives.

The strength classes for solid timber are actually listed in BS EN 338. This contains all of the physical and mechanical properties necessary for day-to-day structural calculations. Timber is allocated to a strength class on the basis of two aspects: its botanical wood species (or species combination) and its structural quality, which is achieved by applying prescribed strength-grading rules. Both visual and machine (automated, NDT) methods exist.

Commercially, the latter tend to be applicable to high-volume applications in smaller sections, such as those used for buildings. Hence for bridging, visual strength-grading is more likely. For this, national standards such as BS 4978 are still used to fulfil the stipulations given in BS EN 518. Comparable requirements for machine-graded timber and machine grading methods are given in BS EN 519.

4.3.3 Strength classes

Classes C14–C30 cover the softwood species, and can readily be fulfilled by visual strength-grading. Machine strength-grading allocates identified wood species directly to strength classes, which is established by initial type testing. However, only smaller cross-sections are produced in this way. Hardwoods also require different methods of inspection, and there is a separate standard designated BS 5756. They are then assigned to strength classes in a similar way to softwoods, but for distinction, the hardwood strength classes are designated D30–D70. For engineering calculations, BS EN 338 is also required. This gives the characteristic strength and stiffness properties, and also the mean and characteristic density values. For specialist applications such as bridges, it is possible to specify an individual, selected timber (identified by its standard name and the botanical species or species group). This may be necessary to ensure a suitable choice in terms of natural durability or appearance. Where this procedure is followed, design data in the form of unique characteristic values may be obtained from specialist organisations.

4.3.4 Natural durability and treatability

These properties are included in *Tables 4.1*, *4.2* and *4.3*. Here the natural durability classifications are for resistance against fungi, the greatest risk in the United Kingdom and most other parts of northern Europe. Information on resistance to other threats, including termites, is to be found in BS EN 350-2.

For most timbers, the decay resistance varies considerably even within each single species. The principal durability class quoted by most standards refers only to the heartwood of the timber, and it is important to note that in no case should the sapwood of any species be regarded as naturally durable. To ensure that the actual service life matches the design expectations, protective measures are necessary – see Chapter 3. These should not be relied upon alone, but should be coupled with choices of natural durability and/or pressure preservative treatment. In this approach, British Standard/European Standard documents are referenced in order to determine whether a selected timber has sufficient natural durability for the exact environment or 'hazard class' in which it is to be employed.

BS EN 350-2 should be read in conjunction with BS EN 460, which indicates the durability requirements in given hazard classes. The primary contents of BS EN 350-2 are two large tables that list the names and origins of many important softwood and hardwood species, followed by their density, immunity against attack from fungi and insects, and also their treatability, or ease of penetration with preservative treatment. The durability classes for resistance to fungal attack have been developed over many years through a combination of historical records and practical and laboratory tests. Referring to the heartwood only, the designations are:

- Class 1 – very durable
- Class 2 – durable
- Class 3 – moderately durable
- Class 4 – slightly durable
- Class 5 – not durable

For *treatability*, a four-class system is used, with the designations:

- Class 1 – easy to treat
- Class 2 – moderately easy to treat
- Class 3 – difficult to treat
- Class 4 – extremely difficult to treat

These are only approximate guides, and in some cases the specialist suppliers of preservative treatment plant and equipment are able to advise on methods that are capable of treating even class 3 timbers. This has been necessitated by extreme exposure requirements, e.g. immersion in seawater, as well as the presence of termites and other extreme hazards in sites outside the United Kingdom.

4.4 Glulam

Glulam has been in existence for more than 100 years. It is the mainstay of timber engineering, called upon whenever larger section sizes and longer lengths are required. It also has a number of other advantages over solid timber. It can be manufactured in a large variety of useful sections (*Figure 4.6*). Curved, tapered and cambered shapes can also be made. It consists of parallel laminations of solid timber that are taken from pre-selected high-quality sawmill stock, which is then accurately dried, strength-graded, end-jointed and thickness-machined. The prepared laminates are then bonded together under pressure and cured, with subsequent finishing machining operations.

Glulam elements act as single, completely homogeneous structural units – in other words there is no slip between the layers, and no requirement for special calculations relating to the bond lines. Thus glulam members are essentially used in design just as large solid timbers, having the further advantage of uniformity of moisture content throughout the section, with subsequent stability in use. Through finger jointing, and the dispersal achieved by mixing the laminations, the effects of natural strength-reducing features in sawn timber are significantly diminished, hence the characteristic values of readily available softwood glulam are higher than those of comparable solid timber.

Figure 4.6 A: large glulam beams (≥220 mm) are manufactured by laterally bonding laminations and laying these up in a 'brick-bond' pattern – this example is a 240 × 600 mm section, 12 m long.
B: using blanks manufactured in this way, glulam fabricators also produce round sections, in this example 450 mm in diameter and 8 m long.
C: a cruciform section produced by re-bonding laminations in the original factory – this must never be attempted on site!
D: a T-section, built up in a similar way. The X and T examples are both in this case 450 mm in diameter, and 8 m long
Drawing © CJM/TRADA Technology

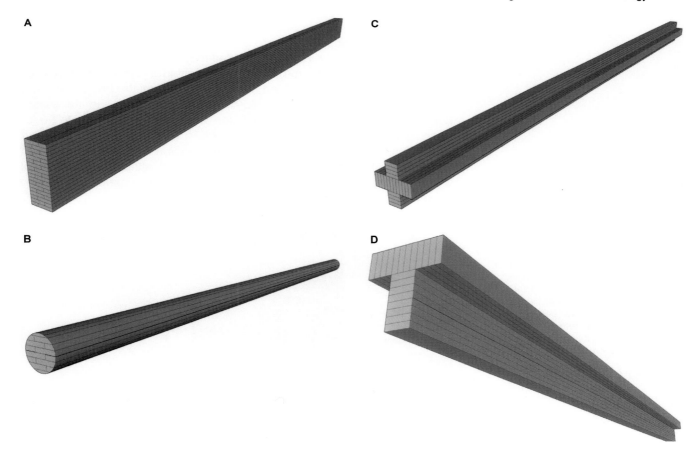

A

C

B

D

4.4.1 Glulam species

The glulam manufacturing process can be applied both to softwood and hardwood laminations, although because of the need to be capable of bonding adequately, only certain species are used. The commonly available sizes of softwood glulam are indicated in the *Manual for Eurocode*, as well as in the individual manufacturers' technical literature. Much of the 'off the shelf' material that is stocked throughout Europe is manufactured from European whitewood, but this is not generally a recommended species for bridging, since it has both a low durability classification and is also resistant to pressure preservation – see Chapter 3.

Specialist timber engineering suppliers can provide better durability in several ways. First, glulam made from European redwood or other species of pine (e.g. Scots pine, Caribbean pine) is more amenable than spruce to pressure preservative treatment (*Figure 4.7*). Another method of enhancing durability is to manufacture glulam from a softwood species having a better natural durability than spruce, and in certain exposure conditions (see Chapter 3) this may make it possible to use the material without pressure treatment. For this approach, the timbers available include larch and Douglas fir – an example of the use of the latter is in the Pont de Crest (case study in Chapter 9). Where this step is taken, it is essential to exclude all of the sapwood from the laminations.

The third means of obtaining glulam with excellent natural durability is to use a suitable hardwood, again with all the sapwood excluded, and with a heartwood rating of durability class 1 or 2. Examples that have an established track record for suitability for bonding and manufacture, as well as for durability in service, include European oak, American white oak and iroko (*Figure 4.8*). In France, there is also experience of laminating structures using idigbo, and there is no reason why this could not also be the case in the United Kingdom. Provided material from plantation sources can be obtained (the only type likely to satisfy 'chain of custody' sustainability assurance requirements), then laminated teak might be considered, operating with a manufacturer who is experienced in the necessary adhesive types and bonding techniques. However, this is likely to be an expensive choice of timber, only suitable for certain clients. In a few interesting cases, the suitability of laminated sweet chestnut has been demonstrated for attractive footbridges (*Figure 1.4*), although it is a material whose supply is somewhat localised, and it requires a manufacturer who is prepared to work with short lengths and small sections.

Figure 4.7 Pedestrian Bridge – Site National Historique de la Résistance en Vercors. Pine glulam, pressure preservative treated with a copper-based formulation; beam edges are covered locally using protective copper alloy. The structural system comprises two simple pairs of bowed beams with a deck passing through. The beams are under-tied with steel, connected to struts that run up between the beams' inner faces. Architects – Groupe 6, Grenoble
Photo © CJM

4.4.2 Durability categories for glulam

Provided the durability characteristics of the adhesive are also considered (these are dealt with separately), the selection approach outlined above, considering wood species in accordance with their natural durability and treatability, is applied equally to glulam.

With these suggestions, it is always essential to select an adhesive that has proven long-term moisture resistance, durability and performance with the timber in question. Generally, this entails a phenol-resorcinol formaldehyde type, which can cope with the acids and extractives often found in these less 'bland' timbers.

4.4.3 Standards for glulam

There is a large number of standards for glulam, but the principal ones are as follows.

The Harmonised Standard is BS EN 14080, giving the details enabling it to comply with BS EN 1995-1-1. Manufacturing requirements are covered in BS EN 386; this includes the weathering properties of the adhesive (*Figures 4.9* and *4.10*). As the second illustration shows, for service class 3 (external) applications, special requirements are included and these are also addressed in BS EN 301. BS EN 390 specifies the tolerances on the sizes, and it also states the reference moisture content at which these are established.

4.4.4 Glulam strength classes

European Standards for glulam do not have a special grading system for the laminations themselves. Instead, a glulam strength class is assigned on the basis of the EN 338 strength class of the laminations from which it is made. Also, the performance of the finger joints in the individual laminations is stipulated, and this may control the characteristic strength of the higher strength classes. BS EN 1194 lists eight glulam strength classes, all designated 'GLn', to distinguish them from the solid timber categories. Within these classifications the material may either be of 'homogeneous' lay-up, meaning that all of the laminations are of the same strength class, or 'combined', where the outer laminations (one-sixth of the depth on both sides of the neutral axis) are of a higher strength class than the inner ones.

The actual allocation of a manufacturer's output to a particular strength class is based on the strength of the species or species combination used

Figure 4.8 Glulam for situations where 'durable' or 'very durable' timber is required. A: a curved iroko glulam component being taken off the formers after bonding using phenol-resorcinol formaldehyde adhesive. B: close-up of iroko glulam Warren girder in an enclosed pedestrian bridge.
Photo A: © Tysons, Liverpool
Photo B: © CJM

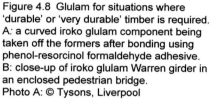

Figure 4.9 The quality-control testing laboratory features strongly in the manufacture of glulam; A: autoclave and shear test machine;
B: oven for cyclic conditioning and moisture testing
Photo © CJM

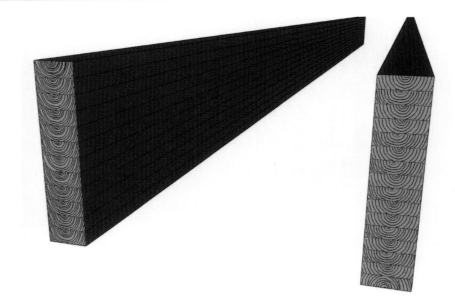

Figure 4.10 Two similar-sized glulam beams, each approximately 540 mm deep and 120 mm in breadth with a length of 10 m. That on the left is formed from 12 45 mm thick laminations, and is suitable for use in service classes 1 and 2. On the right is for service class 3; 16 laminations of 33.3 mm thickness are necessary. This is because BS EN 386 requires glulam in this class to use laminations not more than 35 mm thick. A particular way of laying up the laminations with respect to the growth rings is also demanded – see the difference in the lowest laminations of each beam. Finally, service class 3 material must be manufactured only with Type I adhesives, effectively RF/PRF types
Drawing © CJM/TRADA Technology

as the laminating source. The strength of the individual finger joints in each of the laminations is also considered, and for some types of glulam (the 'c' designations), the position of the lamination grades within the overall beam lay-up is significant. Softwood glulam manufactured in the Nordic countries is often of strength class GL32. This approximates to an earlier common Scandinavian class L40, which is sometimes still mentioned.

Glulam manufactured by other methods, e.g. with alternative lay-ups of the cross-section, may be used structurally in accordance with BS EN 1995-1-1, provided it is an approved product. Also, species other than the commonly available European whitewood may be considered – a particularly important point for bridges since the former is seldom sufficiently durable. Several suppliers offer structurally laminated Siberian larch, for instance, and since this is a more durable species, it is sometimes preferred for external structures or those that may suffer from occasional wetting. Such applications have also been met with durable structurally laminated tropical hardwoods such as iroko, while laminated temperate oak, both from Europe and North America, is occasionally chosen for projects of high architectural merit. Several of these combinations of lay-up and species may lead to alternative strength classes, for which there is provision within BS EN 1194.

4.5 Mechanically laminated timber

Mechanically laminated timber consists of members that are formed from laminations connected together by dowel-type fasteners without any adhesive in the inter-faces. The resulting members may either be straight, lightly curved or cambered (Figure 4.11). The connections between the adjacent layers are normally made with plain stainless steel dowels, with diameters typically ranging from about 9 mm to 38 mm. During manufacture, which takes place in a factory or workshop, the dowels are driven into pre-drilled holes in the laminations, which have diameters smaller than those of the dowels, providing a tight interference fit. This technique evolved from the beginning of the 19th century, and has been used in still-standing British bridges for around 90 years.

Figure 4.11 Middlewood Way Footbridge, near Macclesfield, Cheshire, providing a safe crossing over the A523 for walkers on the Pennine Way long-distance footpath. This is a 50.25 m clear span, tied-arch structure, exemplifying the use of mechanically laminated ekki timbers; built in 1992, recorded on Highways Agency inventory
Photos © CJM

Since around 1950, there has been a revival in Continental Europe, and subsequently in the United Kingdom, in the applications of mechanical laminating. This followed successful applied research and design theories that were provided in German and Dutch universities. Principles and application rules are consequently included in BS EN 1995-1-1. The timbers used are generally dense hardwoods; predominantly West African ekki. However, other species have also been used, particularly European oak.

4.6 Structural timber composites

Broadly, these comprise wood-based panels, such as plywood, and composites produced as longitudinal structural elements. Over approximately the last 40 years, the range of STCs has grown considerably, and at present, the principal types are laminated veneer lumber (LVL), parallel strand lumber (PSL) and laminated strand lumber (LSL). The first of the above is the type most widely used in bridges in Europe, and in this publication only this type is further considered.

4.6.1 Laminated Veneer Lumber (LVL)

The Harmonised Standard referenced by the Eurocode is BS EN 14374. In Europe, most LVL is produced in Finland, under the brand name 'Kerto'. Standard LVL (Kerto-S) has the grain of all of its veneers running longitudinally. This is the common choice for beams and columns. There is a second type (Kerto-Q) with some of the veneers placed across the longitudinal axis of the sheet. This can be used, for example, as the structural substrate of bridge decks, especially in its pressure-preservative-treated form. It is also useful for shear-resistant diaphragms, to form part of a bracing system, providing stability and wind resistance. This commends it for application as a structural roof on modern covered bridges (*Figure 4.12*).

Kerto-Q is assigned separate characteristic mechanical properties, and is referenced through BS EN 14279. Technical advice from Finnforest, the main manufacturer of European LVL, provides characteristic values of mechanical properties; also information on physical properties, standard sizes and fire resistance. The data are assessed independently by the Technical Research Centre of Finland.

Figure 4.12 For the slightly pitched roof of this elegant and ingenious footbridge, Kerto-Q LVL is used to provide wind resistance in conjunction with a lightweight steel system at the entrance portals – Passerelle de Mariac
Photo © J. Anglade

In Australia, LVL is manufactured from radiata pine, on machinery supplied by Raute, the same Finnish engineering company that set up the factories

Figure 4.13 A: multi-layer board press laminated from sap-free Douglas fir, sourced from the same region as the bridge location – Crest, Drôme, France.
B: laminated beech plywood bonded to glulam portals, using full-depth finger-joints – Leimbach, near Zurich, Switzerland
Photos © CJM

Figure 4.14 A: block laminating in progress; the conventional glulam being re-bonded is in the centre of the jig; the wood blocks showing end grain outwards are part of the equipment.
B: a curved-deck pylon bridge for pedestrians and cyclists at Hochstetten, which typifies these new applications
Photos © A. Lawrence & Schaffitzel GmBH

for the European types. With road bridges of a 44 ton load capacity, experience is being gained in building cellular timber decks, as well as simpler stressed laminated orthotropic plates, both combining treated LVL with solid radiata pine elements. These items are illustrated and discussed in Chapter 6.

4.6.2 Innovative STCs

An example of the innovative types of structural material being introduced is the cross-laminated structural panel, whose application in a road bridge deck is illustrated in *Figure 4.13*. Also shown here is a full-depth finger-jointed roof portal for a footbridge in Switzerland. BS EN 1995-1-1, through specification via BS EN 387 makes reference to this technique.

4.6.3 Structural plywood

This is a well-established product that, with care, can be considered in bridge applications such as decks (with suitable protection from water, by sealing) and in the manufacture of roofs for covered structures. Bearing in mind the rigorous demands often occurring in bridge structures, the appropriate types and grades of plywood need to be carefully designated. For compliance with BS EN 1995-1-1, the essential Harmonised Standard is BS EN 13986.

There is a great variety of wood-based panel material used in general construction, and by no means all of it can safely be considered to be 'structural', let alone suitable for use in a bridge. Many types and brands have neither the appropriate quality-control levels nor the associated characteristic values defined under BS EN 1990 and BS EN 1995. Wood-based panels are a complex topic, and reference to documents such as appropriate TRADA wood information sheets is strongly recommended. Special structural plywood products having facings of materials such as epoxy films and metallic envelopes are produced, and these may be particularly useful in some bridge applications.

For plywood, durability depends in part on the inherent resistance of the parent species. However, further factors include the thickness of the veneers; their geometry, arrangement and surface quality; and the properties, quantities and chemical consistency of the adhesives. Further guidance on the specification and uses of wood-based panels in exterior conditions is provided in a TRADA wood information sheet on the subject. Meanwhile, independent-type approvals for design with BS EN 1995-1-1 are available, particularly for materials produced in the Nordic region and in Canada. It is highly recommended that the engineer should make reference to these in the design notes and specifications, then check to ensure that they are actually followed in manufacture, without unauthorised substitutions that may offer lower cost, but lower quality and greater risk.

4.7 Innovations

'Blockverleimung', or block laminating, is an innovative technique recently introduced with success in Germany, where it is regularly used by certain timber engineering manufacturers (*Figure 4.14*). Special waterproof gap-filling adhesives and pressing equipment are essential, and these have been developed with substantial care and investment.

Starting with fairly conventional softwood glulam sections, these form the secondary laminates that are included in a further pressing process. Consequently, very large and almost homogeneous cross-sections become possible. The advantages include very long lengths of component, sophisticated cross-sectional shapes where necessary, and almost complete fire resistance due to the presence of a huge mass of wood. Because of the adaptability of the cross-sections, elements can be designed to resist both horizontal and vertical loads, including extreme point loads where necessary.

Using this method, pedestrian and cycle bridges of over 200 m in length have been installed (*Figure 4.14* and Almere Bridge case study in Chapter 9). The individual 're-laminations' typically start from about six normal strips, making up around 180–240 mm of bonded thickness. Then, pressure-clamping devices with curved heads are used to apply the required forces exactly at the desired positions, with mechanical fastener devices often being added. Test-loading of full-sized structures has been undertaken, as well as monitoring of moisture conditions and tracing the small fissures that develop within the completed material.

Doubly curved profiles can also be manufactured by block laminating, as has been demonstrated recently by prefabricating a 'sculptural' bridge at Sneek, in the Netherlands (*Figure 4.15*). This incorporates a further significant innovation in that acetylated wood has been used as the laminating stock.

This material is chemically modified from permeable softwoods such as radiata pine through pressure treatment using water-based acetic acid. Acetyl groups make the material more durable and less prone to moisture movement by replacing the normal hydroxyl groups within the timber's cell walls.

At present, such projects need to be designed on the basis of bespoke testing and approvals since aspects such as the effect on long-term strength of structural sections is not generically documented. Nevertheless, the attraction of a benign treatment process (effectively the wood is simply injected with vinegar), together with enhanced durability, is likely to drive further applications of a similar nature.

4.8 Metal fasteners, connectors and corrosion protection

In most designs, some fairly significant structural metalware items are necessary, and to ensure a degree of durability that will match that of the bridge timbers, it is necessary to consider their corrosion protection. These

Figure 4.15 A and B: block laminating using 'accoya' – a brand of acetylated wood – for a sculptural bridge at Sneek in the Netherlands. Note the thin strips of laminate that are required to obtain the double-curvature.
C: the Sneek Bridge about to be erected
Photo A: © A. Lawrence
Photos B–C: © F. Miebachn/Schaffitzel

subjects do in principle concern materials. However, since the items are mainly used to form the connections of the bridge, they are covered in more detail in Chapter 7.

4.9 Structural adhesives

The principle of BS EN 1995-1-1 is that the approved structural adhesives have such durability that the integrity of the bond is maintained in the assigned service class throughout the expected life of the structure. The effects of temperature and relative humidity on timber moisture content have a profound influence on the long-term performance of adhesives. Consequently, BS EN 301:1992 makes service class distinctions as follows:

- Adhesives that comply with Type I specification may be used in all service classes.
- Adhesives that comply with Type II specification are only permitted in service classes 1 or 2 and not under prolonged exposure to temperatures in excess of 50°C. *These are generally not recommended for use in manufacturing timber bridges.*

Table 4.4 summarises the options that are likely to make it possible to comply with these recommendations. However, as indicated by the key to the symbols in the table, suitability within the classifications is brand-specific. Hence, a symbol indicating *potential* suitability simply indicates this, rather than the certainty that all brands of the type are acceptable. Consequently, the manufacturer's technical advice, and proof of independent certification regarding use of an intended brand in a given service class and exposure category, is essential.

Table 4.4 Classifications of the family of phenolic and aminoplastic adhesives covered by BS EN 301.

Environment	Adhesive type		
	RF/PRF	MUF	UF
Exterior	+	X	X
≥50°C	+	(+)	X
≥85% r.h.	+	(+)	X
Marine	+	X	X
≤50°C, ≤85% r.h.	+	+	+
Bondline colour	Dark – usually brown/purple	Light – usually white/clear	Light – usually white/clear
BS EN class	301-I	301-I/II	301-II

Notes
+ Generally suitable, provided the specific brand conforms to Type I of the standard.
(+) Some brands may be suitable – manufacturer's proof of certification essential.
X Unsuitable.
RF/PRF – resorcinol-formaldehyde and phenol-resorcinol-formaldehyde; MUF – melamine-urea formaldehyde; UF – urea-formaldehyde.

4.9.1 New structural adhesives

Glulam producers now make extensive use of types of polyurethane adhesive that are not in the above categories. One-component moisture-curing PU types are common, and this practice is beginning to be recognised in European Standards. BS EN 14080 indicates that acceptable strength and durability can be achieved by the use of polyurethane adhesive tested and assessed in accordance with the requirements given in Annex C of the standard. In many cases, such tests have already been conducted, providing the necessary independent certified approvals. For service class 3 exposure, however, the use of phenol-resorcinol-formaldehyde is still necessary, since PU adhesives do not qualify for Type I exposure.

The adhesives used in the block laminating process, described above, also differ somewhat from the standard types used for conventional glulam. They are of the RF/PRF family (see *Table 4.4*) but are specially adapted to tolerate lower bonding pressures and to provide exceptional gap-filling strength.

Research aimed at developing rules for the design of bonded-in rod connections in timber structures has included certain formulations of epoxy resin adhesive. Good results were obtained with these, and the tests included duration-of-load performance and fatigue resistance. However, the adhesive manufacturer's advice remains essential, since the epoxy family is diverse, both in formulation and purpose. Some manufacturers have obtained accreditation for bonded-in connection systems, and the designer should request the necessary proprietary assurance at an early stage.

4.10 Cladding and shielding timbers

Cladding, shielding and louvre components do not necessarily require a design life comparable with that of the principal elements. These items should be detailed so that they can be repaired or replaced with ease. This is discussed in Chapter 3. Nevertheless, maintenance of a decent general appearance to the bridge is a key function (*Figure 4.16*), and the cladding should be selected, detailed and specified so that an attractive durable and lasting external finish is achieved.

Figure 4.16 Horizontal cladding using plain, surfaced four-sided boards that have been pressure preservative treated. In this case, square-edged boards are used, with small spacing gaps between each, for ventilation. Notice the unevenness of colouration due to weathering in exposed and shady zones on the surface. The way to avoid this would be to use a water-repellent microporous and pigmented stain finish. Note also the well-ventilated deck, through which light can be seen. This is another structure of combined glulam, sawn timber and structural steelwork – Pervou Millennium Bridge, near Lausanne
Photo © CJM

Joinery qualities of timber are specified, rather than structural grades, so the designer should follow classifications given in BS 1186-3. Generally, material from class 1 or 2 of this standard is recommended – loose or dead knots definitely need to be excluded, and only quite small knots and relatively straight grain are acceptable. Appearance will also depend on the surface finish – whether planed or sawn, and this relates to the final section sizes too. The choice between sawn or planed surfaced boards affects the arrangement details. With sawn boards, overlapping is normal, whereas louvres or precisely located single-layer gapped boarding are more likely to require planed sections. For more information and building cladding details that can be adapted for bridges, see *TRADA External Timber Cladding*.

The moisture content of installed external cladding and louvres in service ranges from around 10% on surfaces facing spring and summer sunshine, right up to about 22% on shaded and rain-exposed faces in the winter. However, a median range is closer to about 13–19% as suggested by BS 1186-3. It is greatly preferable that the material should be installed in a condition matching the lower end of this range since generally slight swelling after installation is preferable to shrinkage. Hence a target for supply specifications of 14% is suggested, and since material of joinery quality is needed, this is easily achievable.

In all species of cladding timber, both softwood and hardwood, temperate and tropical, the presence or absence of sapwood entails an important specification decision. Suppliers will not automatically exclude this zone from the boards unless the specifier designates so, and in interior or partially sheltered structures this normally does not matter. But for bridges, it is highly advisable to be strict about excluding it unless pressure treatment is intended. These comments apply to fully exposed decking boards as well as cladding, although in the latter case suppliers are probably more accustomed to the rigours of the application.

As well as diminishing durability, the presence of sapwood affects appearance. Surface treatments will soak into the more absorbent sapwood zone more easily, and these will weather differently. Pressure preservative treatment permeates more effectively into sapwood, so if its use is part of the protection plan, then sapwood exclusion is unnecessary.

For cladding timbers, density, natural durability and treatability may be taken from the earlier structural data in this chapter – *Tables 4.1, 4.2* and *4.3*. However, movement classes are also important for this application, and advice on this subject, together with recommendations for fixing methods, are to be found in the external cladding guide referenced above.

4.10.1 Heat-treated cladding

Several proprietary processes for thermally treated wood are now becoming well established. Readily available in Europe is 'Thermowood', which uses technology and plant introduced in Finland. Also established is the 'Plato' process, located in the Netherlands and elsewhere. Cladding is recognised as an important application and standard ranges of board profiles are offered. Thermal treatment is an environmentally benign process that decreases hygroscopicity, which is beneficial in applications such as cladding and louvre manufacture. The dimensional stability of treated boards is significantly enhanced, and their durability against fungal attack is improved without the necessity for pressure preservatives.

Minor losses occur in the strength and stiffness properties of the timber, and it becomes slightly more brittle, but the processes have been adjusted so that these effects are not excessive. Colour changes occur in the surface appearance, but these may actually be considered aesthetically beneficial. After weathering, a silvering and lightening occurs, but treated pine and spruce still look significantly darker and more grey-brown than the natural softwood. If desired, normal surface treatments involving water-repellent colour stains can be applied to thermally treated wood, so this is certainly an option that should be considered for cladding on bridges.

There are standard treatment classes for different end uses. The exterior-use class, applied to pine and spruce, involves an average treatment temperature of 212°C. Thermally treated aspen, birch and poplar are also produced, but these are not established in exterior applications. The claimed improvement in resistance to termites is not fully proven, but it is likely that the lower hygroscopicity and other effects described above will deter the insect pests normally encountered in temperate climates.

4.11 Forest certification and chain of custody

For these two important considerations (closely related but not identical), the two main schemes are those operated by the FSC and PEFC. Both are applied in the Nordic and North American regions, but for commonly available softwood glulam and LVL, the coverage by the PEFC is greater than that by the FSC.

Products from Forestry Commission woodlands and plantations in the United Kingdom are readily available to FSC certification, so this includes eminently suitable bridge timbers such as oak, other temperate broad-leafed timbers (e.g. sweet chestnut) and the more durable softwoods such as larch and Douglas fir.

It is also becoming increasingly possible to obtain FSC-certified tropical hardwoods. This removes a perceived obstacle to their use that has existed in the minds of the public and end-consumers for several decades. Once again, it is now possible to focus on their extremely good natural durability and hardness, and their general suitability for external applications such as bridging. As consumers have become aware through other labelling and certification schemes applying to tropical developing countries, such as 'Fairtrade' food and drink products, the acceptance of supplies, under the correct conditions, assists wealth and employment in local economies, making thoughtless boycotting counter-productive.

5 Concept design

5.1 Introduction

Timber bridges assume a variety of structural forms, sizes and functions, and this provides many opportunities. We expand upon crossings for pedestrians, cyclists and riders, with the possibility of accommodating moderate flows of motorised traffic and giving access to emergency vehicles. At the end of this chapter, we discuss the 'Approval in principle' as a project milestone. In Chapter 7, the important aspects relating to the final engineering design will be introduced, and since these also depend on structural decisions on decks, parapets and related superstructure items, these are examined in Chapter 6.

At the stage considered in the current chapter, the alternatives for architectural suitability and functional stability should be thoroughly assessed, recognising the durability aspects already discussed in Chapter 3 and the materials options introduced in Chapter 4. Agreement within the team that a viable design can be provided is essential before proceeding to the aspects in later chapters, because otherwise there will be serious consequences in delays and added costs.

5.2 Applications

As we have already seen, applications for timber bridges are many and varied. Focusing on some of the most significant, *Table 5.1* introduces the broad design considerations for a selection of applications.

5.2.1 Feasibility

The illustrations and case studies throughout this book show that precedents have been established for numerous structural forms. But from the engineer's perspective, the overall dimensions and proposed layouts are the first things to be considered. Will it be possible to shelter part or all of the main structure by placing it below the deck? This will depend on the general site profile. Broad project plans – including approach paths, tracks or roads; their curves and inclines; the immediate access to the structure and the proposals for its supports – have an impact on the structural feasibility of the bridge itself. Such aspects are often determined by the owner's requirements and are likely to be independent of the decision to use timber, steel or a combination of materials.

Figure 5.1 The landscape and cultural environment has an influence on the choice of form. A: traditional trusses for a 1996 road bridge at Stor-Elvdal, in the beautiful Hedmark county of Norway – a rural region of forestry and farming;
B: a modern interpretation using covered glulam Warren trusses and structural steel masts for a recent pedestrian and cyclist bridge on the outskirts of Grenoble, France
Photos © CJM

Table 5.1 Applications and salient outline design considerations

Applications	Salient design considerations
Footpaths, cycle tracks and bridle paths over roads – often providing statutory rights of way	Often also giving emergency access for light motorised traffic – e.g. ambulances, fire services. For general guidance on layouts, plans and the geometry of vertical sections, follow HA BD 29, in the absence of any other client-specific advice
Railway crossings – usually just for pedestrians and cyclists, sometimes including equestrians and light vehicles	Safety of passengers and public on the bridge and safety of the trains below. Replacing unmanned level crossings. In Norway and France, for example, a number of modern timber bridges for rail crossings have successfully been installed. Techniques such as formal risk analysis procedures have been introduced to generate reassurance on the durability and robustness of timber
Footpaths, cycle tracks and access routes across water features, for example in areas of outstanding natural beauty, national parks, nature reserves and sites of special scientific interest	The terrain is quite varied – it may involve crossing flat-lands (fens, polder or estuaries for instance) or steep ravines or wide valleys. Compared with steel or reinforced concrete, timber weathers particularly well. Agents normally conducive to deterioration are less of an aggravation in a salt-laden atmosphere, typical of seaside and marine situations. For example, slipperiness due to mould growths is less of a problem than on the decks of inland structures
Recreational and leisure activities – very varied applications often favouring timber – includes stately homes and gardens; garden festivals; parks; botanical gardens; golf courses; and seaside attractions	Carrying pedestrians, cyclists and possibly small vehicles for maintenance, general visitors or disabled persons. Bridges in leisure areas will frequently, but not always, be in relatively gentle terrain, but public safety and loading demands may not be inconsiderable
Carrying roads over other highways, across land or water features or railways	Similar comments to the above apply, but obviously stronger and stiffer structures are required – often with more advanced deck systems – see Chapter 6

The influences on choice include not just the apparent distance to be spanned, but the nature and profile of the site, the surrounding terrain and the required clearances and set-backs for the sub-structure. The arrangements of possible abutments, piers and maybe piles and gabions will affect the preliminary planning of the superstructure.

Early in the scheme, delivery and erection considerations also need to be taken into account. As timber engineering is a prefabricated factory or workshop process, it is necessary to consider how the complete structure is to be sub-divided, transported and lifted, generally incorporating substantial part-assemblies. Therefore, the identification of separate components and site assembly nodes becomes another significant concept design factor.

5.3 The structural essentials

Figure 5.2 shows a large through-trussed bridge, capable of carrying vehicular traffic. Such a bridge comprises beams, trusses or arches; transoms and longerons; plus the deck, and other important elements, including the complete parapet structure and various forms of bracing. The structure indicated is broadly similar to one of the spans of the Norwegian bridge at Evenstad, depicted in *Figure 5.1A*. This type has been chosen for discussion at this stage, because it includes the majority of features and design considerations necessary in most timber bridges. In smaller structures, some of these features become less of a challenge. An important point to note about this generic drawing is that most of the aspects indicated are equally applicable to a through-arch structure, although they may be dealt with in a slightly different manner. In Chapter 9, the case study for the truss-arches of the new Tynset arched bridge discusses further important points about this type.

Table 5.2 Features of Figure 5.2

Item	Structural essentials
1 Span	The span range for a framed structure rather than just solid beams or slabs is from approximately 9 m to 70 m
2 Clearance	The clearance beneath the bridge is a function of its purpose, the terrain and, for a multi-span structure, the desired proportions of the whole
3 Upper chords	Upper chords, in this case of glulam, with multiple embedded steel gusset plates to form the connections; note also the protective covers over these vital elements, in this case comprising copper alloy caps
4 Internal webs	Also of pressure-treated glulam, with multiple embedded steel plates and dowels
5 End structures	It is essential to have a means of transferring the horizontal wind forces that affect the structure into the foundations; also, tall trusses and arches require lateral stability bracing and again this has to be taken into the ground. Various methods are possible; here, the end-structures are steel portals. The impact resistance of the end structures also needs consideration
6 Deck	In this case, a stressed laminated deck is used. Other options include simple planked beams; decks built over transoms and stringers; glulam slabs, T-sections or hollows; nailed laminated decks; and timber–concrete composites. Because of this wide range and its importance, deck design is covered separately in Chapter 6
7 Lower chords	Here, the stressed laminated deck acts as a slab, so including the function of providing lateral stiffness to the lower chords, but in general, this aspect, as well as the protective design of these vital elements, needs careful planning
8 Tie-hangers	Tensile members of round cross-section, from high-strength corrosion-protected steel, are supported beneath the upper chords; at the lower ends, multiple steel fins pass through slots in the lower chords, connecting onto the steel transoms
9 Transoms	Hanging from the upper chords (see above), these support the stressed laminated deck plates. They also form a stiff lateral base for the containment parapets that are, in this case, a combination of steel and glulam
10 Upper bracing	A solution is required to stabilise the trusses or arches, at the same time giving adequate headroom for bridge traffic. In this case, 'K-bracing' is used down to the point at which a free space is required between the trusses. Note that the overhead timber braces also require a protection scheme, following the principles in Chapter 3
11 Parapet structure and guard rails	Parapets have to contain people or vehicles for safety, but they have a significant aesthetic and 'user-experience' effect, too. They must provide a sense of security, but often, especially with timber bridges that are chosen because of their pleasantness in the environment, they need to allow users to see the landscape and to contribute to the visual profile from a distance. Their contribution to the durability of the bridge should be considered right at the start of the concept. Further details of parapets are discussed in Chapter 6

Figure 5.2 A through-trussed timber bridge showing its main features and considerations, most of which are equally applicable to a through-arch structure. The key to these features with an outline of their essentials is given in *Table 5.2*
Drawing © CJM/TRADA Technology

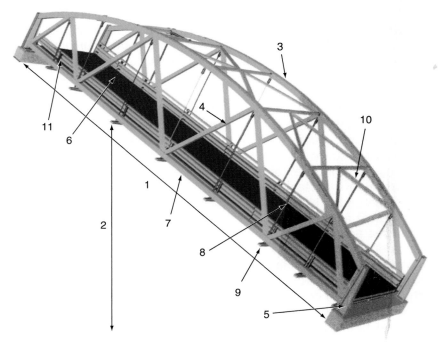

5.3.1 Arrangement decisions

Using multiple dowel-type connections, very large forces can be carried, and this is regarded as one of the major advances of late 20th-century timber engineering. Use of these affects the primary arrangements significantly. In horizontal or lightly cambered flat girders, truss components and latticed arches, chords and web members are placed in single planes and member centre lines can be located to intersect, with near-ideal exactness.

Assuming one requires a span beyond a simple beam or slab, then the general lightness and uniformity of loadings on most pedestrian bridges often qualifies these for consideration as arches. Of course the topography and site context need to be appropriate, but when this is so, the arch is one of the most efficient ways of transmitting compressive forces caused by self-weight and live loads. Sometimes the arch may beneficially be tied, in order to avoid transmitting large thrusts into the abutments.

With large point loads and high bending moments, arches tend not to perform so well, so an alternative is to use trusses. Although the latticed truss bridge was pre-eminently a 19th-century solution, as we saw in Chapter 2, with modern timber engineering techniques, the main components – the trusses themselves – can largely be prefabricated in the factory and transported to site in their completed condition. Again, single-plane arrangements are made wherever possible, using inserted plates and multiple steel dowels. These were the choices with the bridge shown in *Figure 5.1A* and in *Figure 5.2*. Alternatively, smaller assemblies such as prefabricated truss panels are possible, which can rapidly be linked together before launching.

With truss arrangements, it is usually possible to ensure that the live loads are effectively applied at the nodes. The global bending moments form a couple, with compression forces in the top chords and tensile forces below. Similarly, the shear forces can be globally considered as acting through the diagonal web members, in tension and compression depending upon the configuration. Because of the guidance on the load duration factors given by the timber codes, including the National Annex to BS EN 1995-2, it is usually possible to make a simplifying preliminary evaluation of resistances. This may make the assumption that the variable actions due to the passage of vehicular and pedestrian traffic are of short-term duration, being sufficient to gain an approximate indication of likely member sizes and choices of configuration, materials types and strength classes.

5.3.2 Prefabrication and erection

During the early project stages, transport and erection provisions influence the design, affecting the shaping of the elements and the manufacture of the connections – particularly the erection nodes. The identification of the site erection nodes and the division of major components into vehicle loads or packs for delivery should never be overlooked. Although the techniques of prefabrication and launching or lifting are essentially similar to those for steel structures, the lighter mass of timber gives greater scope for the use of cranes, winches, temporary towers and other similar methods. The option of division of prefabricated and trial-assembled elements into smaller parts permits delivery to the most remote of sites (*Figure 5.3*), but in general, the preference is to maximise workshop assembly, where the conditions are dry and controlled, with stable lifting equipment quickly available.

Figure 5.3 Man-liftable prefabricated modular timber bridge panels in Chile.
A: being unloaded;
B: assembled into requisite number of under-deck girders according to span and loading;
C: temporary bracing being added to girder after lifting and prior to decking
Photos © CJM

It is wrong to think of timber engineering as a process principally involving on-site building that begins with simple 'sticks' and uses hand-held tools. Glulam structural elements, for example, can be manufactured in very large sizes and a great variety of shapes, but this work needs to be completed in a factory or workshop. Dependent upon the chosen protection plan, there may be additional pressure preservative treatment processes, and these should be carried out on components that are almost completely fabricated and with nearly all carpentry operations completed. This makes it possible for preservative to be injected into the more vulnerable end-grain sections, where slots, holes and recesses have been cut.

Transport solutions are diverse. As well as road delivery, water transport, railway delivery and even helicopter lifting are known (*Figure 5.4*). The environmental costs of delivery can now be measured scientifically – see the case study for Crest Bridge, in Chapter 9. In this respect, water transport and rail delivery are significantly more benign than road delivery. However, the influence of this factor on the final choice will depend on the distance between the fabrication location and the site, as well as the immediate approaches to the site, where offloading and erection are critical factors.

Because of the relative lightness of timber components (*Figure 5.5*) – generally in the order of 850 kg/m³ including allowance for steelwork – vehicle load capacity is rarely an issue. The additional influence of preservative treatment on density may need to be considered, and such figures should of course be verified for every individual project. The shape of non-linear components such as arches is also important in selecting delivery methods, identifying erection nodes and determining costs. While not large, the space occupied on transport by steelwork attached before delivery may affect the packing of loads. During delivery, supports need to be provided to curved shapes to avoid distortion and over-stressing, and temporary protection against dirt and mechanical damage will also affect the shape and bulk of loads.

The Flisa Bridge, Chapter 9, is a good example of a successful design involving prefabrication. It was erected in 2003 in a space of only three months, and opened on time, only seven months after closure of the steel

Figure 5.4 Lightweight timber components make helicopter delivery feasible in very steep terrain
Photo © STEP

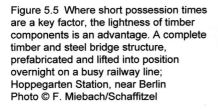

Figure 5.5 Where short possession times are a key factor, the lightness of timber components is an advantage. A complete timber and steel bridge structure, prefabricated and lifted into position overnight on a busy railway line; Hoppegarten Station, near Berlin
Photo © F. Miebach/Schaffitzel

structure that it replaced. Its individual truss chord sections were restricted to about 28 m for delivery, but some of the erection nodes were designed as stiff connections, rather than pins. Factory treatment was applied to the nearly finished and pre-shaped parts, meaning the lengths that could be inserted in the cylinders were a factor in deciding the break-down of the prefabrication.

5.4 Architecture

In designing a bridge, its purpose, the general environment and site, as well as the approach routes, have a significant influence on the architectural and engineering decisions.

5.4.1 Form

Even though it is strongly dependent on the planning factors outlined above, the form is fundamental to the desired architectural statement. Sometimes a bridge may be sought which provides a landmark structure. In other instances, the brief may be to blend into the landscape, or to complement the surrounding architecture. The overall impression needs to relate to the context, whether this is a natural landscape, a town or city (*Figure 5.1*). Protective design measures for timber bridges, a feature discussed in Chapter 3, are strongly reflected in the architecture, and as already mentioned, should not be treated as 'add ons' or afterthoughts. When well conceived, these essential features make a positive contribution to the aesthetic statement, as well as practically ensuring an extended life.

5.4.2 Aesthetics

These need to be considered both from a distance and in close-up (*Figure 5.6*). For those using a pedestrian bridge, the crossing itself should be a pleasant experience and rather than expecting people to pass by in a hurry, the designer should give attention to the details, encouraging them to linger and enjoy the experience. Thought needs to be given to the textures and colours of all the materials, not just when they are completely new, but also after a reasonable passage of time, given the maintenance that can be foreseen.

The form selected for the bridge generates its visual expression of efficiency, its order and unity and its artistic shaping. An aesthetics checklist for form, adapted from 'The appearance of bridges and other highway structures' is shown in *Table 5.3*.

Figure 5.6 The graceful arches and elegant roof of the Leimbach Footbridge, 32.4 m span, blending into the soft landscape, over the river Sihl, Canton Zürich, Switzerland. In close-up, the faces of the arches and piers are recessed. This is an aesthetic tribute to the famous Bernese bridge engineer, Robert Maillart, 1872–1940, who constructed inspiring reinforced concrete bridges in the region
Photos © CJM

Table 5.3 Aesthetic principles

Visual expression of efficiency	Order and unity	Artistic shaping
Slenderness	Organisation of structural system	Visual expression of flow of forces
Transparency through openings	Coherent cross-sectional shapes	Cross-sections that minimise stresses
Mass and balance	–	Light and shadow effects
Form relating clearly to function	–	Structural and non-structural ornamentation

A bridge that is generally regarded as ugly can usually be explained by relating the functionality to the aesthetics, for most people have an instinctive appreciation of structure, even with no formal training. Trusses should be arranged as transparently as possible and their nodes should visually fulfil their function. Diagonal bars and ties should relate to one another with geometrical consistency. Cluttered members should be avoided, since they add visual 'weight' to the structure. Consideration should thus be given to using steel for some of the internal ties, for instance (*Figure 5.7*). The cross-sections of members should be kept consistent with one another, again both for technical reasons and for visual appeal.

5.4.3 Material textures, finishes and details

The revealed surfaces affect the final appearance, but they must be specified in accordance with the durability strategy discussed in Chapter 3. Unless protective finishes are applied, the surfaces that are evident to bridge users will alter significantly with time. In making the necessary choices, the ambience of the bridge is important, although somewhat subjective. Selecting textures and finishes affects appearance at three levels:

1. The broad surfaces themselves, both visually and on a tactile level.
2. The appearance of the junctions between materials.
3. The mix of materials in the complete structure.

The decision as to whether to specify artificially coloured finishes relates to durability as well as to the newly built aesthetics. Bare wood exposed to sunlight and the elements soon changes significantly in appearance, often for the worse. Even on naturally durable timbers, microporous water-repellent wood stains are highly recommended – see, for example, the Pont de Merle (*Figure 3.19*). These systems have undergone decades of testing and development, and their use deters fine checks on the surfaces. These later lead to algal growth and unsightly stains, if not encouraging wood-destroying decay.

Both translucent and opaque types of finish are available. They may represent the natural colour of various species of newly prepared wood, or other shades across the full spectrum can be selected according to taste and possibly fitting local custom or planning requirements. Self-coloured pressure preservative treatments may be an alternative. In texture, a great variety and scale of depths and coarseness have been developed in all of the principal wood finishes, but choosing, for instance, a rough-sawn surface may be precluded by the more precise dimensional demands of an engineering design.

Figure 5.7 Very high flood levels occur periodically in the town of Mariac, as can be seen by the mark on the ancient stone arch bridge, A. To provide a pedestrian connection between the Mairie and a public recreational area across the river, the architect provided this elegant lightweight solution with a timber deck suspended from glulam posts that support an LVL roof (B, C). An 'inverted suspension' catenary helps to hang the deck from the roof. Light steel end posts take the lateral forces down to the foundations. Consequently there is no structure below the deck level
Photos © J. Anglade

Careful consideration of the junctions between various materials contributes to the overall aesthetic success, as well as generally improving durability. The aim should be to express the behaviour of the connection or boundary, respecting the nature of each of the materials involved. Simplicity is the key to success; a complex detail usually means unresolved problems. Breaks between disparate materials should be visually clean, and as crisp as possible.

The variety of materials in the overall design should generally be restricted, and where the generic class of material changes, the reason should be read distinctly and logically, even to the layperson's eye. Often the use of massive materials such as masonry or concrete may be appropriate for the supports. These may then successfully be combined with superstructures involving the more linear and slender materials, such as steel or timber. Such decisions need to appear logical and, as with the close detailing, they should be bold and decisive.

5.5 Terrain

The nature of the terrain and the local contours often dictate choices. Examples of different sites that all invoke particular structural forms are:

- road or rail crossings over cuttings, built-up or altered ground;
- waterway bridges in flat-lands – fens or polder for instance – or in steep ravines or wide valleys;
- bridges in leisure areas will frequently, but not always be in relatively gentle terrain;
- city or docklands sites often entail existing harbour walls, embankments and established routes.

Amongst the important factors that will be influenced by these primary features are the estimated costs of delivery and erection of the superstructure. For crossings on road, rail or water networks, statutory requirements covering, for example, safety and clearance criteria are often vital. These are obtained from relevant bodies such as the Highways Agency, local authorities, Network Rail, the Environment Agency and British Waterways.

For river crossings, the flood levels advised by national or regional river authorities must be taken into account. The precise location, abutment and soffit levels in relation to the flood plain will need to be detailed. For some watercourses, the potential influence of erosion may also have a critical impact on the choice of bridge location.

Bridges spanning road or rail networks may require steps or ramps, and the site will be assessed to decide how to provide a sufficient bridge approach area to construct these.

5.6 Geology and site access

The foundation conditions at the proposed site will be determined following the advice of qualified ground engineers. The accessibility of the site during construction and its influence on the bridge design and project costs should be considered. Difficulties in transporting large structural members to remote sites and the likely costs of using heavy plant and machinery for

Figure 5.8 Examples of features influencing the general arrangements. A: high limestone cliffs provided the opportunity to site massive abutments for the tension anchors; diagonal strutted timber piers lift the tension-ribbon to give clearance for laden barges and boats' masts.
B: narrow winding roads, steep rock faces and deep gullies – plenty of headroom beneath the deck for an under-strutted frame, but delivery of compact components for easy erection is essential
Photo A: © 'Peter'/Flickr
Photo B: © TRADA Technology

Figure 5.9 A combined-use bridge for pedestrians and cyclists in Wroclaw, Poland. This structure is cantilevered from one bank, with V-section tapering block laminated timber beams and an illuminated glass parapet
Photos © Schmees & Lühn and their architects

this purpose may limit the choice of the bridge structural form. The means of carrying out any heavy foundation or approach path work that may be required should also be considered. In considering road and rail crossings, possession of the carriageway or track will be another important issue.

As indicated in Section 5.4 above, costs can be minimised by limiting work at the bridge site, and carefully studying alternatives to reduce transportation and lifting costs.

5.7 Location, purpose and functional geometry

The precise location of the crossing should ideally lead to the shortest possible bridge span. However, other considerations may influence this choice, such as the orientation of the approach paths, site features, substructure and abutment costs. In some locations the approach route may be difficult to alter. For instance, surrounding land, buildings or statutory boundaries may restrict the positions of the footpaths or cycle-ways on either side of the bridge.

For vertical and horizontal clearances for footbridges over highways in the United Kingdom, reference is made to documents published by the Highways Agency (HA). Section 6.1 of HA BD 29/04 deals specifically with these points, also making reference to Departmental Standard TD 27, *Cross-sections and Headroom*. For vertical and horizontal clearances to footbridges over railways, canals and other watercourses, the appropriate authority will similarly indicate outline profile standards. In the case of structures following the HA requirements, the horizontal clearance from the edge of the carriageway to the bridge supports is normally a minimum of 4.5 m, but where this cannot be achieved, there are strict collision requirements for the supports.

For bridges intended purely for pedestrians, width requirements in BD 29 are based in part on the peak pedestrian flows, with an absolute minimum of 2 m (see *Table 5.4*). Separately, Section 12 of BD 29 gives the requirements for combined uses involving pedestrians, cyclists and equestrians. The summary for these is included in *Table 5.4*. Such shared facilities may be segregated or unsegregated. *Figure 5.9* shows an unsegregated footbridge shared by pedestrians and cyclists in the city of Wroclaw, Poland. In the United Kingdom, cycling both as a fitness sport and as a means of commuting tends nowadays to entail quite high-speed pedal traffic, so operating organisations tend to prefer combined-use bridges to contain at least a degree of separation. As an alternative to railings between different users, there may be white lines, or colour-contrasting pigmented non-slip strips within the different zones of decking. The latter are also encouraged as a means of improving accessibility for visually impaired users.

Crossing bridges involving equestrians (normally in a mixed-use situation), may well entail heavier design loads than mixes of the previous two traffic types. Indeed, BD 29 calls for the use of another HA Standard, BD 37, which invokes highway loads for horse traffic. This also raises the issue of motorised traffic on what are essentially footbridges or combined cyclist–pedestrian types. To deter the risk of this happening in a unauthorised manner, BD 29 Section 3.15 gives guidance on barriers and bollards.

Table 5.4 Key dimensional requirements according to HA Standard BD/04.

Purpose	Minimum useable width of bridge deck, ramps and stairs	Height of parapet from upper deck surface to handrail
Pedestrians only	2000 mm	1150 mm (1500 mm over railway)
Pedestrians and cyclists	2000 mm unsegregated 3000 mm with line, colour contrast or texture separation 3500 mm with a separate kerb 3900 mm with separating railings	1400 mm (1500 mm over railway). On ramps, consider inclusion of chicane barriers to slow mounted cyclists, but allow also for mobility aids – see BD 29 Clause 12.6
Pedestrians and equestrians	3500 mm	1800 mm in all cases

Sometimes, particularly in premises operated by limited liability or private organisations with permissive use, there may be the intention to allow controlled access for emergency vehicles, in which case the bridge engineers will consult with relevant overseeing and operating authorities.

5.7.1 Parapets

In the types of use where BD 29 or similar standards apply, bridge spans, ramps and stairs are all required to have parapets with significant conformity rules of their own. Those of the greatest significance are summarised in *Table 5.5*, but the designer will no doubt understand the need to consult further, since there are many measures that are not timber-specific. Examples of parapet and handrail designs that do pertain to this are given in Chapter 6. BD 29 Section 7 is specifically for this topic, and the HA Standard provides further references to various DMRB documents, national and harmonised European Standards.

Equestrian bridges may entail special, high in-fill panels to obviate the risk of animals being startled by traffic on the road or railway under the crossing.

For architectural distinction, and sometimes in conservation zones, for instance, laminated glass parapets are considered. These are recognised by BD 29 and other approvers, with strict requirements for post-fracture strength and shard behaviour. These are now being used in conjunction with contemporary timber structures as shown, for example, in *Figure 5.12*.

5.7.2 Enclosed footbridges

This publication contains a number of illustrated examples of fully enclosed footbridges – *Figures 5.7* and *5.10* show two. BD 29 contains Section 8 on this topic, mentioning several situations where the additional security provided by this type of crossing is advantageous. These include where

Table 5.5 Further dimensional requirements affecting parapets and handrails according to HA Standards

Parapet, posts and in-fill	Handrails	Ramp/deck gradients
Heights as in *Table 5.4*; vertical in-fill with a maximum gap of 110 mm; height of opening below deck surface and lowest rail 100 mm maximum; post spacing 2000 mm max	Between 900 mm and 1000 mm above ramp or deck surface; diameter of 50–60 mm suggested	Generally not steeper than 1:20; there may be consultation with client or overseeing organisation regarding steeper localised lengths of deck/ramp; mobility requirements are often determined in consultation with local access and disabled user groups

Figure 5.10 Passerelle de Vaires-sur-Marne. The timber solution was chosen taking into account the constraints of a very short possession time, the safety aspects of the launch and the ease of transporting the components from the manufacturer
Photos © Jacques Mossot/Structurae

Figure 5.11 Views of railway footbridge structure – span 50 m; width of deck 9 m; height 10 m. The finished volume of Douglas fir glulam is 300 m³ while for the deck, some 20 m³ of American white oak is used
Photos © Jacques Mossot, Structurae

there is an exceptional risk from objects being thrown or dropped off the bridge, and where there is an unusually high history of persons jumping onto the carriageway or railway. Other motives include sheltering users at sites exposed to very adverse weather or where the deck is so high that pedestrians might otherwise feel insecure. This section of BD 29 contains its own specific guidelines on geometry and clearances for such bridges.

5.8 Rapid construction for a railway: Passerelle de Vaires-sur-Marne

This railway-crossing footbridge comprises three parallel frames of tied-arches. Lateral stability is ensured using secondary cross-arches that are connected to posts that rise from the principal tie beams, supporting a roof. Hence, to some extent, the construction resembles the historic Burr arch-trusses.

Completed in 2004, an outstanding feature of the project was the rapidity of construction, once all the prefabricated Douglas fir glulam components had been delivered to the assembly yard. At a busy SNCF station east of Paris, a bridge had to be provided overnight, to span tracks for a new TGV-est (the high-speed train service) extension. All of the components were prefabricated at a plant in Alsace, while the assembly site for connecting together the final parts of the structure was about 300 m away from the bridge's eventual position across the tracks. Access for two 24-wheel cranes each of 700 tonnes capacity entailed some strengthening of approach crossings.

The finished volume of Douglas fir glulam is 300 m³, while for the deck, some 20 m³ of American white oak is used.

For the actual launch, the contractors were permitted track closure between 22:00 on a Saturday night and 06:00 on the following morning. The launch was completed with 30 minutes to spare, and public access at the station was provided within one month of erection.

The project aims have been totally fulfilled. Outstanding features amongst these are:

- For the people of Vaires-sur-Marne, safe pedestrian access to town areas behind the railway tracks – previously some fatal accidents had occurred at a level crossing.
- For SNCF station users, a pleasant crossing to the quays for trains to Paris and Province.
- For the directorate of SNCF, a convincing demonstration by the architects and engineers of the competence and merits of timber bridges – in general, timber structures like these have not been used on French railways for some 100 years.

5.9 Choices of structural form

The following five categories of structural form have already been introduced in Chapter 2:

- beams, including bowed and under-tied types with no significant arch action;
- cantilevers;

- suspended types – considering cable-stayed and suspension bridges, and also pure tension-ribbons;
- arches – in this chapter we need to consider both free and tied-arches;
- trusses – also including parallel girders.

These main choices of form are summarised in *Tables 5.6* and *5.7*. First, these tables consider their main static systems, features and options; second, they review the alternatives for deck elevation in conjunction with the main static arrangement. Following the tables, each of the principal types is briefly discussed and illustrated in terms of concept design. Finally, a short section considers moving bridges – lift and swing types – which can adopt various structural systems.

These static system options cannot be considered in isolation. As we have seen, the functional geometry arising from the purpose of the bridge and its traffic types is a strong indicator. The terrain, site topography, ground conditions, approach routes and slopes also dictate or eliminate choices. The various structural options also interrelate with durability, an obvious example being that if all the other factors allow a straight choice, then trusses or arches below the deck are preferable to those above it. In conjunction with the summary tables, Chapter 3 gives more information on durability, and Chapter 4 on materials. Chapter 6 will expand on decks and parapet structures, while Chapter 7 will consider engineering design.

5.9.1 Deck elevations
There are fundamentally three options, with a number of variations and constraints. For the forms shown in *Table 5.6*, there may be low-level or under-slung decks, or mid-level or high-level decks (*Table 5.7*).

The choice of deck level has considerable influence on the architectural shape and distant aesthetics, as well as affecting the engineering design. However, as with the choices of static systems in *Table 5.6*, this is strongly dependent on planning considerations and primary functional aspects, including the acceptable geometry of the general arrangement.

5.10 Beams

The beam bridge is the simplest, albeit rather restricted form – because of its limited spanning potential. But under the right circumstances it may provide the most effective and economical solution, and it need not be unimaginative (*Figure 5.12*). *Table 5.4* further summarises the static systems, span ranges and main features of beam bridges. In principle, no engineer should have difficulties with the concept design of these.

Even during preliminary design, quick checks for probable adequate strength in shear (mainly parallel to the grain) and bearing should be included, since both properties relate to potentially weaker resistances in

Figure 5.12 A simple laminated beam bridge – designed in softwood glulam with an open deck of durable hardwood. Metal covers (as described in Chapter 3) protect all the below-deck elements, and a water-repellent stain finish is applied. This structure is of 14.2 m span, and typically just over 2 m deck width. With the protective measures, the durability is good, and such bridges are extremely economical
Drawing © CJM/TRADA Technology

Table 5.6 The main choices of form

Structural form	Beams and cantilevers	Suspended types	Arches	Girders and trusses
Static systems	Single simply supported span; single span cantilevered from both abutments; Vierendeel girders	Simply supported main span, cable-stayed; simply supported span, suspended; single tension-ribbon tied at each abutment	Circular, parabolic and elliptical; three-pinned; two-pinned; tied; latticed arches (trussed internals)	Horizontal or lightly cambered girders – Warren, modified Warren or Pratt formation – all three over-deck or under-deck; Town and Howe forms still occasionally used; bow-string trusses; A-frames with low-slung deck; kingposts forming triangular side profile
Alternative forms	Slight bow, but not arched; box sections from glulam (suitably ventilated); single-spanning slab, i.e. with plate action; multiple simply supported spans; continuous spans; cantilevered side spans with suspended central section; strutted and under-tied	Single or twin masts or towers; tilted or cranked pylons; curved or winding deck with masts offset/tilted to accommodate lateral forces; multiple span tension-ribbons with intermediate hinges and propped by braced triangulated struts (Essing) or steel 'trees' (Ronneburg)	Multiple spans; arched over-deck with low rise – creates high thrusts; arched stressed laminated decks acting as two-way spans	Multiple spans achieved in various ways – commonly just repeated simple spans with intermediates on piers, e.g. Vihantasalmi – but cantilever arrangements also exist (e.g. Flisa, 180 m with 71 m main span). Lenticular trusses (underslung). Trusses – often Warrens – arranged as triangular-sectioned frames
Shape of principal members	Straight; lightly cambered; tapered; alternative sections, e.g. triangular	Straight or lightly cambered beams; tapered beams; curved cross-sections, e.g. inverted aerofoil; concave cambered beams for tension-ribbons	Curved profiles by glue laminating or mechanical laminating; for tied-arches, steel or glulam members; near-arch action using straight sections to form polygonal frames	Straight or lightly cambered chords for girders; more pronounced curves for through-trusses, e.g. Evenstad
Common materials – see Chapter 4	Sawn timber – softwood or hardwood; glulam of many shapes and sections; round-section timber – natural but turned or turned glulam; stressed laminated slabs – flat or cambered in side elevation; mechanically laminated timber – usually hardwood.	Flexural members as for beams; block laminated timber decks for curved plans; masts in glulam; hollow glulam, e.g. from polygonal sections; masts in specialist hardwoods, e.g. basralocus; steel masts and pylons	Taper (in elevation or plan) easily achieved with glulam; large timber arches, e.g. Tynset – 70 m – have latticed arches with bi-tapered glulam chords and glulam webs. Mechanically laminated timber – usually hardwood – e.g. Middlewood Way, Macclesfield	Sawn timber – softwood or hardwood; glulam of suitable shapes and sections; round-section glulam – e.g. Traunreut; mechanically laminated timber – usually hardwood

timber. Also, it should be borne in mind that while timber is a high strength-to-weight ratio material, its stiffness does not increase at such a rate with elevation of strength class or species type. Hence static serviceability design procedures are important, and general guidance on the subject, aimed at timber engineers, is well worth attention.

Beams range from a single, simply supported span to multiple spans and cantilever arrangements. In laminated construction, pre-cambered and slightly bowed forms (without expressly designed arch action) are quite common. Span ranges for beam bridges may be from as little as 3 m for a very small solid timber footbridge to about 24 m in bowed laminated construction. One of the simplest forms of bridge is the constant-depth beam type.

Table 5.7 Deck elevations and variations.

Structural form	Beams and cantilevers	Suspended types	Arches	Girders and trusses
Deck elevations	Deck over beams; deck between beams – near centre line or close to bottom edges, dependent on choice of deck framing; two-way spanning slab, e.g. using stressed laminating (slab = deck), allowing narrow side elevation, similarly achieved with block laminated decks	Generally similar to beams	Over the crown of the arch; between the arch-frames; suspended from the arch	Similar options to those for beams and cantilevers – decks between framed girders commonly use U-frame action to stabilise upper chords
Additions and variations	A roof, often used to assist lateral stability and wind resistance as well as offering shelter	Under-slung lenticular trussed frames – e.g. former Traversina Footbridge	Roofed arches; Burr arch-trusses with stairs – Passerelle de Vaires; skewed plan forms – Dutton Horse Bridge	Roofed through-girders; including triangulated sections

Table 5.8 Beam bridges, structural systems and typical spans

System	Typical span range	Features
1 Simply supported beams, straight or lightly bowed; transoms, stringers (can be lightly bent), cross-boarded deck	4 m (solid timber) to 28 m (glulam)	Economical, simple structures for short spans; good potential for protection of beams using an over-deck
2 Through-beams, superstructure as above or, e.g. flat, cross laminated plank deck; treated LVL deck; solid nail laminated deck; glulam slabs; timber–concrete composite deck	12 m to 26 m – usually glulam beams	Economical; attractive side-elevations; span–load capacity limited by possibility of only one pair of beams; lateral–torsional buckling check method given in Eurocode 5.
3 Under-strutted beams, using braced triangulated struts and/or rigid sub-frames; decks as in system 2.	7 m (solid hardwood) to 29 m (glulam).	Structurally economical; protective top-decks as in systems 1 and 2; requires suitable site
4 Continuous beams with intermediate piers; either moment connections or central span 'hung' off side spans that are cantilevered out from piers; central span optionally slightly bowed	18 m to 42 m central span (glulam); decks as in systems 1 and 2, or, e.g. Crest, Drôme, France, has MLB deck	Overall, gives possibility of quite a long bridge, e.g. *Figure 5.15*
5 Under-strutted beams with steel ties – effectively simple trusses	Normally about 40 m max.	The former (1996–2001) Traversina Footbridge was very lightweight and reached 47 m span. This type is only suitable where there is the essential deep clearance beneath

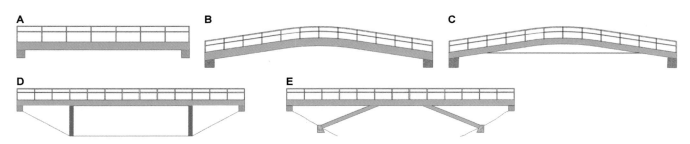

Figure 5.13 A: simple straight beams;
B: simple lightly bowed beams;
C: bowed and under-tied with steel;
D: continuous spans with intermediate supports from piers or timber struts;
E: understrutted with rigid frames, upper beams continuous over struts, optionally pinned at centre
Drawing © CJM/TRADA Technology

Figure 5.14 A five-span constant-depth beam bridge – note that since an open-boarded deck is used, the upper edges of the glulam beams are protected. The balustrade is moment-connected by means of inserted steel fins. Built at Gschröff by a paper mill, part of which is still working and also housing a museum of the industry Photo © F. Miebach Scaffitzel

Figure 5.15 Triple, tapered glulam beams for the cantilevered footbridge in Wroclaw, Poland – this is the structure also illustrated in *Figure 5.9*
Photos © Schmees and Lühn

Figure 5.16 A slab bridge formed from sets of straight glulam members, stress laminated to achieve plate action. The span of this example is 14 200 mm, with a deck width of 3000 mm, designed for light axle loads. Protection is through a combination of oil-based pressure preservative for the deck, protective design features (sloping surfaces, air gaps, cover sheathing) and a water-repellent stain finish on the superstructures. This bridge has a triple-layer sealed deck, and it uses the detachable parapet described in *Figure 6.27*. For schemes where this type of bridge may be suitable, a similar effect with higher load-carrying capacity is achieved with the cellular stress laminated deck shown in *Figure 6.10*
Drawing © CJM/TRADA Technology

In the superstructure of a beam footbridge, the slenderness ratio (span–depth) of the main beams is generally between 20 and 30. If the outer spans are designed as continuous over the supports, then shallower span-to-depth ratios can often be achieved (*Figures 5.14* and *5.15*). This may also lead to fewer expansion joints, which are often a source of maintenance problems due to water penetration, requiring excessive attention to the joints themselves, and also causing the staining of piers and abutments.

The number of beams and the spacings naturally affects the slenderness ratio, and there is scope for ingenuity (*Figures 5.11* and *5.16*). The beams should be arranged so that access for inspection and maintenance is possible. Design rules address requirements for lateral diaphragms and cross-frames, whose presence influences stability, load distribution and serviceability performance. While these may be thought of as 'technical' issues, the presence of such transverse elements needs to be well planned, since for low-level observers it has a strong influence on aesthetics. The soffit view is an important and often forgotten 'elevation'.

5.10.1 Two-way spanning slabs

Slab bridges in concrete are common, and these have a corresponding form in timber. One of the first 'modern' types was the stressed laminated deck. The variations in these are considered in more detail in Chapter 6.

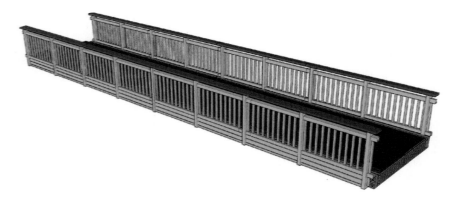

Another quite recent alternative, principally developed in Germany, is the block laminated type (*Figure 5.17*). In both cases, the deck is designed as a diaphragm, with two-way spanning plate action, and the consequent appearance in side elevation is quite slim.

The way the edges of decks are detailed and how the parapets and the fascia are designed strongly affects the appearance of the bridge. Careful attention to the form and detailing of any cross girders or transoms that the structure requires ensures that the appearance of the bridge from beneath will not compromise its overall aesthetics.

Figure 5.17 A block laminated deck being installed – the Jagstbrücke, at Möckmühl, Germany. The structural system includes cable stays, for which one of the support pins is in the foreground
Photo © F. Miebach/Schaffitzel

5.11 Cable-stayed, suspension and tension-ribbon bridges

The broad principle of the cable-stayed and suspension bridge, *Figure 5.19*, is clearly that the deck is directly supported from elevated masts or towers, via pure tensile ligaments.

Cable-stayed structures (*Figure 5.18*) are nowadays more common than catenary suspension bridges, which were regarded as more suited to very long spans. However, in remote mountainous regions such as the Himalaya, modern engineering principles have continued to fulfil bridging needs by using steel and timber, where previously natural fibres or iron chains would have been employed. In classical, wild and romantic landscapes, such as forest parks, large estates and gardens, small classical suspension bridges for pedestrian use may continue to fulfil a valid role (*Figure 5.19A*).

Figure 5.18 An impressive cable-stayed bridge at Czorstyn, Poland
Photo © Josef Schmees

An advantage of the cable-supported structure is that the span can be sub-divided with aesthetic clarity, which allows the deck to become slender and visually light. Just as with a simply supported beam system, however, the beam and deck structure does require thorough stabilisation, ensuring it is resistant to any tendency to buckle, and proof against wind forces. In the structural zone where the braced timber structure is connected to the foundations, steel framing arrangements may be used to assist (*Figure 5.20*).

Figure 5.19 A: Capillano Suspension Bridge, near Vancouver – a tourist attraction, originally built in the 19th century with hemp ropes, but maintained and updated with modern materials, included a graded timber deck;
B: pedestrian road crossing of 24 m at Balliagues, Switzerland, using turned round timber for the masts and main beams.
Photo A: © Wikimedia Commons
Photo B: © STEP

Figure 5.20 The cable-stayed footbridge at Hochstetten comprises a block laminated deck with strong, sweeping plan curvature, introducing further complexities into the design engineering, but as can be seen, the resulting slenderness of profile makes this very worthwhile
Photo © F. Miebach/Schaffitzel

The cables and masts themselves have a definitive visual impact, which generates drama and can lead to very striking structures. Tilted pylons are common, as represented in several of the illustrations, and in general, it is preferable to have a clear single or double plane of cables. With a fan arrangement, the two planes of cables are often tilted inwards, which both assists technical design and also resolves itself visually. Modern manufacturing techniques render possible the use of very shallow profiles with sweeping plan curvature (*Figure 5.20*). These types involve significant engineering challenges in the analysis and design of wind effects and bracing.

Towers or masts should be designed to give an impact that is compatible with the configuration of the cables and transoms. Anchorages should not be overwhelming, and should visually express their function of transferring forces. Each tower or mast should appear to rise from the ground or water in a unified fashion, right to the top. Often they are swept forward (*Figure 5.18* and *Figure 5.20A*), which again assists the engineering while complementing the appearance. With the more traditional cable-stayed examples in *Figure 5.21*, it can be seen that raised approach embankments have been necessary to allow barge or ship clearance beneath the deck.

5.11.1 Tension-ribbons
Until recently, Dietrich and Brüninghoff's Main-Danube Channel footbridge at Essing was the only well-known exemplar in timber, but recently the same team installed another impressive timber tension-ribbon at a garden festival in Northern Germany (*Figure 5.22*).

A

A

B

B

Figure 5.21 A: a more traditional cable-stayed example using timber towers – Tuibrug Enschede, Netherlands – mechanically laminated ekki beams and deck, with braced towers of basralocus.
B: glulam beams and deck incorporate a viewing platform – Bereldange, Luxembourg
Photo A: © CJM
Photo B: © Schmees & Lühn

Figure 5.22 The 'Drachenschwanz' tension-ribbon bridge, built for the 2007 Garden Show at Ronneburg, Germany.
A: one of the block laminated segments during delivery – note the 14 inserted steel pin-plates to form true structural hinges;
B: the completed structure in use
Photos © F. Miebach/Schaffitzel

These structures look daring, and in the appropriate context they have a dramatic visual effect. In forested surroundings, for example, they reflect the natural trees in the landscape. For the 2007 Garden Show at Ronneburg, Germany, one of the important targets was permanently to reclaim and settle a large area of land that had been scarred by mineral extraction.

While apparently simple, tension-ribbons require sophisticated design and construction. Their dynamic behaviour, for example, requires considerable attention. Also, substantial anchorage is necessary (*Figure 5.23*), and this type of bridge can only be considered where it is possible for it to be incorporated logically within the topography.

5.12 Arches

Site, terrain, ground conditions and clearance considerations may lead to the choice of this form, which is architecturally very commanding. Much larger spans are possible than with beams, a range in the order of 12–70 m being feasible. Within the general family of arches, various deck arrangements and positioning levels are possible (*Figure 5.24*). Choice amongst these also depends of course upon the terrain and other general features.

Architecturally, the arch frames and their deck need to be considered as an entity, in conjunction with other general arrangement requirements such as structural stability. The significant horizontal thrusts generated by the arches have to be resisted either by tying (*Figure 5.25*), or by abutments, and in order to rely on the latter, suitable ground conditions are essential.

5.12.1 General arrangements

As indicated above, several alternatives for deck level may be possible, dependent upon the requirements for the profile and ground/water conditions as well as the required clearances. With a high-level deck, and arches situated beneath, there is greater shelter for these primary elements, and scope for increasing their number beyond a single pair. A tied-arch bridge involves a through-deck, requiring ingenuity in bracing design in conjunction with clearances for traffic. The tied-arch is, in

Figure 5.23 As well as the innovative block laminated deck, a significant challenge for the 'Drachenschwanz' Bridge was the design and construction of the foundations and abutments
Photo © F. Miebach/Schaffitzel

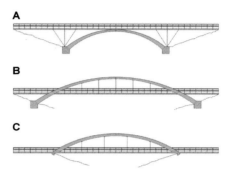

Figure 5.24 Arch types, including aspects affecting their choice of deck level and associated bracing systems. A: arches with a high-level deck – road bridges of up to 45 m span have been built as this type. B: in less steep terrain, arches with an intermediate deck are convenient; they are often considered for crossings over main highways, motorways and railways and can carry both foot and motorised traffic. Spans of up to about 50 m are feasible. C: where the headroom is less demanding, low-decked arches may be preferred; latticed types in this form have spanned up to 70 m with road traffic
Drawing © CJM/TRADA Technology

Figure 5.25 A pair of tied-arches at Guben, Germany – commonly a slender plan is required for a footbridge, while with a low deck such as this, clearance is required beneath any transverse arch bracing; consequently, inward-canted frames may assist in achieving lateral stability; note also the copious protective design features
Photo © Josef Schmees

Figure 5.26 'Rainbow arch' – generally unacceptable except in landscapes, because of its steep inclines
Drawing © CJM/TRADA Technology

principle, little different to the bowstring truss bridge, a well-established form that is still successfully employed under the right circumstances. Tied-arches and bowstrings are chosen where almost complete prefabrication of the timber structure is desired, possibly for remote sites, difficult ground conditions or risks of adverse weather.

An intermediate deck level has neither the advantage of providing shelter to the arches, nor such facility for minimising site work. However, it may be chosen for an elegant profile, and to match certain approach gradients and traffic clearances. For footbridges, cyclist and equestrian traffic, canted arches, popularised by bridge architects working with steel, have also been built in timber. Their conceptual design is similar to that for the vertically arched, suspended deck, but with more complex analytical demands against wind loads, bracing performance and in-plane deck forces. For a roofed bridge, low-level decks, and more rarely those at an intermediate height on the arch, can be considered, as discussed in Chapter 2.

Also only appropriate for certain sites and profile parameters is the very low rise-to-profile bridge with a lightly bowed over-deck. The alternative option of approach steps, used historically in, for example Chinese 'rainbow' bridges (Chapter 2), is now generally only appropriate in landscape gardens or similar sites, where its steep inclines are less of an objection. At the other extreme, the flat arch brings challenges to the engineer in terms of providing acceptable heights of vertical cross-section, as well as in resisting its pronounced horizontal thrusts.

The structural mechanics of timber bridge arches are no different to those for steel structures. Both two-hinged and three-hinged arches are common, but there are reasons for preferring one or the other within individual projects. Either type may be tied if necessary. Transporting the prefabrication imposes limits of about 30 m span, and so large arches may be delivered as a pair of components with a pinned hinge at the crown. Moment-rigid crowns have also been used successfully, so this is not an automatic reason for the choice. Where ground conditions are softer or less reliable, the three-hinged type tends to be regarded as more tolerant of potential support displacements.

5.12.2 Arch geometry
Parabolic, circular and, occasionally, elliptical arches are all produced. The parabolic arch (or theoretically the inverted catenary – not quite the same), gives the purest compression system, with the influence line for the action effect following the shape of the profile. This is only true for uniform loading, however. Also, to carry a deck, the arches must either receive braced columns, with a high deck, or support hangers, with a low one. In either case, these verticals introduce concentrated loads on the arch itself. Together with the unsymmetrical deck loadings that arise in practice, moment effects therefore have to be provided for.

5.12.3 Bracing
Free-standing arched frames, as occurring in a bridge, require carefully designed and properly connected bracing to provide lateral restraint to the arches, and to resist and transmit wind loads. This is discussed in more detail in Chapter 7, but here we can note that both timber systems and steelwork are used. In a longer span and heavier-loaded bridge, the

bracing itself may have to carry significant forces, mainly in an axial direction. Round-hollow steel sections are sometimes selected, giving a light and unobtrusive impression, as well as being easy to connect and maintain.

In connection with the level of the deck, the method of stabilising the arches requires careful planning, and traffic clearance may well conflict with the ideal positioning. This particularly applies to the suspended and intermediate deck types. A high-rise, top-decked arch bridge can have the main bracing at evenly spaced nodes right down to the support surfaces, but only certain site profiles will accept such a bridge.

Even though longitudinal temperature changes in timber are negligible, care must be taken over similar effects that may unsettle the structure in the longer term, especially for longer spans and lower rises. Through creep, possibly exacerbated by moisture cycling, dimensional changes and alterations in deck geometry may become cumulatively significant over a large arch span.

5.12.4 Cross-sections

The classic glulam arch has a cross-section that is effectively made of solid material, because although it is manufactured from bonded laminates, normal timber engineering calculations treat it as a generic substance, without the necessity for special checking at the bond lines. For small or moderate spans such solid curved glulam arches have quite sufficient load–span capacity. However, the manufacturing process makes it difficult to provide breadths of section greater than around 215–240 mm. Therefore, very large curved glulam arch components are sometimes designed by using two closely paired members in each arch cross-section. These are individually of up to 240 mm in width, and of the requisite depth, but with the adjacent 'plies' sandwiched almost face-to-face, and just having packing pieces in between. This method of lateral linking ensures the necessary restraint against torsional buckling that would arise with a single very slender cross-section, at the same time ventilating the zone between the adjacent faces.

For long-span arches, modern latticed constructions have been developed. In the Nordic region these have now been used for some time, finding applications both in road bridges with spans of up to 70 m (Tynset Bridge, Norway). These arches have high, curved top and bottom chords, that in perspective look quite slender in section. Internal, triangularly trussed webs are connected by means of flitched-in ribbed steel plates, connected to the chords with plain steel dowel fasteners. The bolt and plain dowel heads and tips are protected against water penetration by means of recessed timber plugs. Many other protective design features are included in these types of bridge, as described in Chapter 3.

5.13 Trusses

As discussed in Chapter 2, trussed bridges were amongst the earliest of types to have a recognisably 'modern' engineering content. In terms of contemporary trussed road bridges, the Flisa Bridge (Chapter 9) is an outstanding example. Built in timber in preference to steel, as a replacement for an older bridge in the latter material, this bridge holds the current record for the type, with an overall length of 196 m and a main span of 70.3 m.

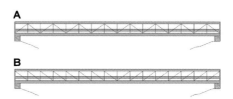

Figure 5.27 Common parallel-chorded truss types; both forms have good span–load carrying capacity and the potential for strongly protective design features. In some situations lack of headroom beneath the structure may be a constraint.
A: modified Warren configuration;
B: Pratt configuration
Drawing © CJM/TRADA Technology

Figure 5.28 A modern Warren trussed covered footbridge at Traunreut, Germany, during erection. Contrary to impressions, all of the red-earth coloured members are of glulam, not steel! In the foreground, some temporary cross-bracing for erection
Photo © F. Miebach/Schaffitzel

One of the major limitations of older timber trusses was the technology of the connections. This has been overcome by using compact connections formed with multiple steel plates that are inserted within the glulam chords and web members. These 'flitched-in' plates are then rigidly connected to the timber using multiple dowel-type fasteners. Such systems are fabricated with CNC techniques and avoid the problems of timber connector bowstring trusses and similar older types, whose connected faces needed large overlap areas, with multiple plies of members giving a cumbersome overall thickness.

Trussed girders provide greater load-carrying capacity and stiffness than simple beams. Various trussing arrangements are possible, including Warren, modified Warren and Pratt. With a fully below-deck structure, girders are often formed from several lines of trusses. These can easily be cross-linked with bracing, without an impediment to the carriageway. The main disadvantage is, of course, that the free height beneath the bridge is reduced, compared with the other common alternative, the deck between the girders and slightly below the neutral axis (*Figure 5.27*). These types of design normally involve other lateral members, such as transoms, which may be of timber or steel. For bridges of the type shown in *Figure 5.27*, bracing of the upper chords is often achieved with 'U frame' action. This is discussed in Chapter 7.

Figure 5.29 A lightly bowed parallel-chorded girder truss bridge, using the Pratt formation – internals comprise longer ties, in steel, and vertical struts in glulam. Footbridge at Düsseldorf
Photo © Schmees & Lühn

As well as varying deck levels, the introduction of a camber or light curvature may be considered (*Figure 5.26*). Well-designed timber girder bridges are architecturally pleasing – viewers are able to 'read' the structural forms, and appropriate designs can be conceived for both urban and rural situations. Individual spans for bridges formed from girders of this type are likely to range from about 9 m to 45 m.

Modern timber engineering versions of several traditional timber bridge forms have also appeared recently (*Figure 5.28*). Both 'bowstring' and 'kingpost truss' types have been given an updated treatment through the use of new connection technologies, innovative deck types and environmentally sensitive timber treatment processes. Use of these forms has been extended into multiple spans creating some of the longest timber bridges constructed in modern times.

Figure 5.30 A modified warren trussed girder bridge of approximately 30 m span, in this case including a light-duty stress laminated deck – a type of design suitable for pedestrian and mixed light traffic applications. Girders may be of the form shown here, or of other classic truss types such as the Pratt configuration
Drawing © CJM/TRADA Technology

Figure 5.31 A pair of kingpost trussed cycle track bridges, one each side of the main road – in Seoul, South Korea
Photo © Schmees & Lühn

5.14 Summary: the main aspects of a successful timber bridge

The protection strategy, the principal components and elements, and the main timber items with their commonly chosen materials are summarised below.

Define the protection strategy – there must be one!

Either

- choose a *very durable* species;
- use treated timber – with moderate coverings; or
- provide thorough covering and cladding/screening with a less durable species.

Always – ensure good drainage and ventilation!

Clarify the principal components and elements:
- main beams;
- chords and internals;
- arches;
- deck – structural functions; strength, serviceability; type of deck structure; see Chapter 6;
- cross-beams or transoms;
- parapet items;
- all connections;
- wind and stability bracing and means of transmitting lateral forces to the foundations;
- deck and parapet bracing.

Designate the main material choices. Commonly they are:
- glulam – for the principal components;
- dowel-type fasteners and inserted steel plates for connections;
- steel for hangers, tension rods, parts of bracing, sometimes parapets and transoms;

- solid timber, glulam or LVL for stressed laminated decks;
- block laminated timber decks for curved plans and sections – requiring contact with specialist producers.

5.14.1 Principal components

It is usual to select glulam for all the main beams, transoms, chords and internals of girders, trusses, arches and similar forms. The outstanding advantages are clearly its availability in long lengths and large sections, with flexible shape options. Its uniformly low moisture content throughout is maintained by the protective design measures discussed in Chapter 3. Both timber and steel options may be considered for transoms, parapets and bracing parts. Either glulam or steel transoms can be tapered – use of the latter may be necessary for vehicular bridges, due to the need for a much higher shear capacity.

5.14.2 Deck design

A major topic in its own right, the conception of the deck is addressed in Chapter 6. Although this aspect needs to be linked with the overall structural form, most types of timber deck can be given preliminary consideration at least, in any scheme. And indeed timber decks are often chosen over steel-framed bridges also. Clearly the heavier point and line loads induced in a motorised traffic bridge are likely to require some form of laminated deck, or a timber–concrete composite structure. For a purely pedestrian bridge or for the paths of a road bridge, an open-boarded deck, or a simple cross laminated type may be adequate. The role of the deck in protecting the main structure should not be overlooked, but this has already been discussed above and in Chapter 3.

Further elements:
- copper or zinc-alloy protection covers;
- corrosion-protected structural steel for hangers, tension rods, bracing, parapets, etc.;
- corrosion-protected bright plain carbon steel or stainless steel for dowel-type fasteners;
- durable, high-quality thin-section timber or treated wood for cladding items – boards and louvres;
- open decks – gap-boarded with durable timber non-slip sections – typically castellated hardwood;
- sealed decks – stressed laminated most common; other options are discussed in Chapter 6;
- associated deck items – membranes, high-performance asphalt systems, seals, wearing surfaces.

5.14.3 Other functions

In connections, the dowels transfer the lateral shear between the timber and inserted steel plates. This should always be detailed so that there are open slots to permit drainage. Sealed deck plates are designed specifically to act as diaphragms, and Sections 5 and 6 of BS EN 1995-2 were written especially to facilitate them.

High-level arches and through-trusses require tie rods to hang the deck, as well as a means of dealing with the zone where lateral bracing cannot be included for headroom reasons. In the lower connections, the inclusion of fixity between the tie rods and steel transoms is one method that can be used to assist stabilisation. Increasing member breadth can enhance lateral stiffness where the arches or truss chords drop to the upper foundations. This method can be in conjunction with pinned supports to the bearings that can also tolerate lateral forces.

End blocks or pillars are often required for impact resistance, and these may be of steel (which can easily be portalised) or of reinforced concrete. The complete parapet needs to include items that protect bridge users and any under-traffic that exists. They may contribute stability to other parts of the structure. For vehicular bridges, kerbs and crash guards are normally necessary. These need not necessarily be entirely of steel – combined steel–timber systems have been designed, crash tested and put to use in existing bridges.

5.15 Obtaining outline approval for the design

The Black Dog Halt SUSTRANS Bridge (sustainable transport – cycling initiative) shown in *Figure 5.32* crosses the A34 in Wiltshire, England, at the location of a former railway branch line that is now a cycling route. It is an example of the type of bridge for which an 'Approval in principle' (AIP) will formally be sought before the final engineering design work is undertaken. The term 'AIP' is one used by the UK HA, but in most projects elsewhere, for other clients and approving organisations, there will be a similar milestone. As we shall see in Chapter 7, even the 'final design' is usually staged, and consulting engineers often organise projects to distinguish between several phases within this, ending at quite a late stage with complete validation calculations, approval and the freezing of drawings and specifications ready for fabrication and erection.

Figure 5.32 'Black Dog Halt' is a former branch line station on a SUSTRANS route (leisure and fitness cycling initiative). Because of the alignment of the former railway, the structure crosses a dual carriageway with a curved and skewed plan
Photo © CJM

It has been the aim of this chapter to bring the reader to the point of AIP. In cases where approving authorities other than the HA are involved, then a similar formal break-point to AIP is highly advisable, separating the concept design work from the full calculations and details. This is normal in most construction projects, but because of the great variety that is possible in timber, it is especially significant that such agreement should be reached so that major and expensive alterations are unnecessary once the final design work starts. Even for a private client, this permits the assessment of alternatives for architectural and functional suitability, as well as durability and robustness. Costs are far more likely to be controlled when a viable design is agreed before committing to lengthy analyses, detail drawings and calculations.

6 Decks and parapets

6.1 Introduction

Decks are of paramount importance, in addition to the main load-carrying spanning elements, such as the beams and arches, discussed in Chapter 5, since without them it is difficult to cross the bridge! Timber decks can be built in a large variety of sizes, loading capacities and performance capabilities (*Figure 6.1*). Hence, this chapter is devoted to them, although as already mentioned, they must be brought into focus at quite an early stage, being primary structures and forming part of the overall bridge concept design.

Various further superstructure items – kerbs, handrails and balusters, for instance, are usually attached more or less directly to the deck, and these can conveniently be taken into account at this stage. We also sometimes need to consider ramps and steps or external stairs made from timber,

Figure 6.1 A: a simple open-boarded deck, using pressure preservative treated Scots pine; note also the handrail posts and parapet extended from flitched-in steelwork sandwiched between the glulam carriage beams – Vercors, France.
B: a more sophisticated sealed deck using modern technology – this is a block laminated deck, using a 220 mm deep layer of glulam, followed by a membrane, then a 40 mm plate of Kerto LVL, sealed with a bitumen felt and a two-layer mastic asphalt system; in case of partial breakdown of the waterproofing, there are drainage collection points and downpipes – a true 'belt and braces' system – Tharandt Botanical Gardens walkway – see case study in Chapter 9
Photo A: © CJM
Photo B: © Schmees & Lühn

and for convenience these too are included. In general, the roofs of fully covered bridges follow the main principles for those on other buildings. However, some of the shapes and coverings that are especially useful for bridges are shown, together with guidance on the necessary shelter angles at the eaves, and some brief discussion of how the roof may assist the overall stability of the bridge.

6.2 General functions

The deck supports the traffic – a general term that includes pedestrians, cyclists and equestrians, as well as powered vehicles. Often, the actual structure of the deck does not provide the direct running surface, but it is the substrate upon which waterproof membranes, sealing layers and wearing surfaces are placed (*Figure 6.1B*). The running surface is then designed to provide grip and to resist wear, transferring the concentrated loads of the traffic onto the main deck structure, and subsequently through to the main load-bearing elements.

The deck must be proportioned to resist all the applied loads arriving from various directions, and it must meet serviceability design criteria. These include static deflection limits and possibly further considerations such as the avoidance of excessive vertical and horizontal vibrations. The formal calculation approaches for satisfying these requirements are addressed in Chapter 7. In addition to its primary vertical load-carrying role, a deck is often designed as a horizontal diaphragm. Acting compositely with the main structure, it may thus enhance both the vertical and the longitudinal load-carrying capacity. As a diaphragm, the deck may also brace the main structure and transfer horizontal actions into the principal linear elements, i.e. the beams, trusses, arches, etc.

6.2.1 Protection

As we saw in Chapter 5, the deck often acts as a roof or protection for the main structure, which may be located below or surrounding it. However, the scope and facility to create a general arrangement of this nature depends on the site factors that we reviewed previously. If it is necessary to employ tall trusses or arches with overhead bracing (*Figure 5.2*, for instance), then alternative methods have to be found to protect these vital members. To improve protection either of the deck itself, or of other parts in addition, connections to the deck may be made to support lateral cladding. Wherever possible, it is worth trying to ensure that the deck effectively shelters the remaining structure from moisture and the delayed effects of precipitation. Both of the timber–concrete composite decks shown in *Figures 6.16* and *6.17*, create significant shelter for the timber structures beneath them. As we discussed in Chapter 3, this will beneficially effect the definition in terms of hazard, or 'use classes' of the underlying structures.

6.3 Solid timber decks

Spaced sawn planks may be used to form simple one-way spanning, non-diaphragm decks. They are usually laid transversely, but are sometimes placed longitudinally or in a herringbone pattern. Longitudinal plank arrangements may appear to be efficient structurally, but are less convenient for cyclists and other wheeled users. Two- or three-layer plank decks are also possible, but since these are effectively a simple form of cross laminated deck, we consider them below.

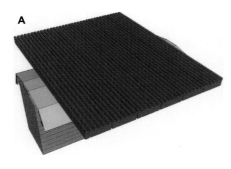

Figure 6.2 A: an open-boarded hardwood deck on a softwood glulam longeron – because the latter is not a 'very durable' species, it has protective covers, and these need to be added to all parts of the below-deck structure (e.g. transoms, blocking), which is also coated with a water-repellent stain finish.
B: a simple cross-braced glulam beam footbridge showing the recommended support method for transverse, spaced, sawn planks – these are mounted on elastomeric strips that lift them off the beam top edges. There is a longitudinal runner supported half way across the span of the transoms, which reduces the deck board span. Note also copper alloy protection over all the below-deck elements, which are also coated with a coloured water-repellent stain finish
Drawing © CJM/TRADA Technology

Solid timber decks may comprise softwood or hardwood, with the more dense hardwoods preferred for maximum wear and durability. Structurally, the planks are specified using BS EN 338 (see Chapter 4). Suitable pressure-preservative-treated softwoods include Scots pine or European redwood; radiata pine may equally be chosen in countries where it is readily available. Untreated Douglas fir or larch may also be acceptable, provided it can be ensured that it is completely sap-free.

With a solid timber deck, the designer should face the important fundamental choice between an open-boarded type or a sealed superstructure. The latter is only advisable where one can be confident that reliance can be placed on future maintenance. A high quality of construction is required, such that the water will always drain off the deck and never onto the structure below. For this reason, sealed decks are usually within the domain of more sophisticated structures such as the block laminated deck already seen in *Figure 6.1*. Note that this structure includes the 'belt and braces' principle.

Figure 6.2A shows the recommended arrangements for a spaced sawn plank deck. Deliberate gaps are included between the boards, and double protection – covers plus a water-repellent stain finish – is provided for the structure below. Here a durable hardwood deck is used over softwood glulam longerons. If a 'very durable' timber such as mechanically laminated ekki is used for the main under-structure, then it is possible to omit the covers and optionally to leave out the surface treatment, although the latter is still a good idea to prevent initiation of surface checking. Even with the 'very durable' species option, it also remains good practice to mount the deck boards on longitudinal bearing strips of a hard elastomeric material (*Figure 6.2B*). This both improves user comfort and reduces moisture trapping. Hardwood decking planks machined from various naturally durable species (*Figure 6.3*) are commercially readily available, and are also used extensively in timber decking situations beyond footbridges. Because of this substantial market, various forms of 'anti-slip' insert strips are routinely provided. These are generally formed from granite, bauxite or other hard mineral particles, which are blended together and bonded in place using epoxy adhesives, inserting them in recesses in each board. It is possible to purchase prefabricated planks containing these strips coloured with dyes in red or green, enabling them to be used to demarcate zones, e.g. between cyclists and pedestrians.

Amongst the few durable hardwoods occurring in European woodlands, sweet chestnut is a good choice for decking boards, provided that the available dimensions and quality are suitable. 'Heartwood only' planks of European oak are also possible, but it should be borne in mind that this species has high movement characteristics, and needs to be selected for quality as well as for structural grade. Certain temperate hardwoods of the southern hemisphere, such as rauli, are rated as 'durable', unlike their European cousins, the true beech. Consequently, they may be specified for decking where their availability indicates so.

Amongst the most common commercially available tropical species are iroko, jarrah and ekki. Also available are greenheart and purpleheart – these are very durable and hard, as well as being virtually 'antiseptic' to moulds that cause slipperiness. Both have a long and successful history of use for decks and heavy-duty industrial flooring.

Some lesser-known but durable tropical hardwoods are being introduced as available with FSC chain-of-custody certification, and this includes some species that are not listed in European Codes or Standards. Specialist organisations may be consulted to obtain design properties for timbers such as cumaru, ipé and massaranduba. Technically these appear to be entirely suitable, provided their quality is assured.

All of the above-mentioned hardwoods can be specified through BS EN 338, selected for strength to the HS grade of BS 5756 or an equivalent in other national grading standards. As indicated above, expert advice may be necessary to provide unpublished classifications. This leads to identification within the 'D Class' EN system of mechanical properties, referenced by the Eurocodes. In this respect, communication between the designer and supplier is important since the latter may not always be fully conversant with the procedures in relation to the standards. In the cases mentioned above, where commercially available pre-prepared decking is marketed, suppliers often issue indicative span tables. Provided the engineer can confirm such calculations, this is likely to be a good route.

In public areas where decking is used, profiled decking planks containing castellated grooves (typical groove size 8 mm width × 8 mm depth) are now widely specified because of concern over litigation after slipping accidents. A 'Stanley pendulum test' – BS 7976: 2002 Part 1–3 – is used to qualify slip-resistant decking, which also often contains the aggregate strips mentioned above.

Timber Decking, published by TRADA Technology, is a useful guide to decking design.

Since decking is exposed to abrasion, the engineer should make an allowance in the strength and stiffness calculations for a loss of section. The maximum depth of wear before maintenance replacement should be known, but this ideal state of affairs is unusual. Edge grain planks are more resistant to wear than flat grain ones, but it is not usually possible to stipulate the method of conversion so exactly. Where flat grain planks are used, they should be attached with the surface that was closer to the heart of the tree placed uppermost, so that in the event of cupping, water is shed rather than being retained. Edges should be provided with a radius, so that water is drawn off by surface tension. In rural footbridges, the gaps between planks should not be less than 6 mm, so that dirt and debris can pass through, allowing air to circulate. For urban or suburban situations where users often wear lighter shoes, a gap of 4 mm is typically recommended.

6.3.1 Laminated veneer lumber

Techniques to treat LVL that is manufactured from rotary peeled spruce veneers have meant that this material may be considered to be sufficiently durable to use in bridge decks, provided it is well protected with sealed membranes and wearing surfaces.

An LVL deck may be used as a solid plate (*Figure 6.4*). Effectively this becomes an alternative to the solid timber deck, where similar principles apply with regard to protective design, drainage and sealing options. Either the all-parallel veneer type of LVL may be used, or the cross-veneered type can be chosen. The second type will offer improved lateral stiffness and plate action – these aspects are discussed in Chapter 4. LVL is also employed in built-up, stressed laminated decks. These are considered in this chapter.

Figure 6.3 An open-boarded deck in Monmouth, using castellated-section ekki, fixed with countersunk stainless steel screws to make replacement easy
Photo © CJM

Figure 6.4 A pressure preservative treated Kerto LVL deck, covered by proprietary metal non-slip decking and an under-membrane (not visible) – Black Dog Halt SUSTRANS Bridge, near Calne, Wiltshire
Photo © CJM

6.4 Laminated decks

Capable of carrying heavy vehicular traffic, as well as being suitable for lighter loadings, these have been a significant development in timber bridge engineering. Several general forms have been developed to provide longer spans and to carry heavier loads than are possible with simple planked or boarded surfaces. All of these offer more effective ways of collecting and distributing concentrated loads. When taken in conjunction with more efficient sealing and protection methods, and better detailing to avoid water trapping, a greatly superior and more durable structure results, in comparison with an open-boarded deck.

Apart from the cross-laminated plank decks, the simplest and probably oldest type of built-up deck uses vertically arranged boards acting similarly to joists, but held in close contact with one another by means of a precisely designed dense nailing pattern. In nailed laminated decks, the connection to the main structure is usually also by nailing, although more sophisticated forms of attachment can be used, avoiding the necessity for skew-nailing.

After nailed laminated decks, simple types of glued laminated slab evolved, and like the former, these are also still a potential choice, being designated by BS EN 1995-2. For glulam slab decks, proprietary types of shear-locking attachment have been developed to facilitate positive and easily assembled connections to the sub-structure.

Pre-stressed timber deck plates were first produced in Ontario, Canada, in the 1970s, as a means of upgrading nailed laminated decks. In North America and Australia they are now common, while in Europe their use is growing, encouraged by their inclusion in the Eurocodes. In Austria, Germany and Switzerland, they have been used for some time, and Nordic designers have also adopted the system – see the Norwegian road bridge case studies in Chapter 9, for example.

Decks that are pre-stressed but not completely bonded together (unlike full-width glulam slabs) are documented in several bridge design codes, including that of Ontario itself, and subsequently that of Australia and the timber Eurocode. These systems employ transverse tendons of high-strength steel to clamp the assembly, providing significant orthotropic plate action. At the edges, special members are introduced to resist the high bearing stresses perpendicular to the grain of the timber. These edge strips are either from hardwood, or from U-section steel, while sometimes a combination of the two is used, together with spaced clamping plates.

6.5 Cross laminated plank decks

Unlike the types described later, the cross laminated plank deck just uses boards placed on the flat, and in several layers (*Figure 6.5*), with various forms of alternating grain arrangement. For example, the adjacent layers may cross one another at 90°, or at 60°/30° (for an approximately isotropic plate effect using three layers). Each layer is normally simply connected to the next via a similar set of mechanical fasteners to those used to attach the deck to the under-structure. More sophisticated systems are possible – for example rigidising the deck with separate improved nails (corrosion-resistant ring-shanked nails); using special screws; or introducing an inter-layer of waterproof polymeric adhesive.

For these types of deck a simplified analysis approach is provided in BS EN 1995-2. As well as being useful for small, low-cost footbridges, where they may contribute to lateral bracing, they also sometimes find application for the raised footways of larger bridges where there is a separate vehicular carriageway (see, for example, Vihantasalmi Bridge – *Figure 6.19*).

One of the innovations that has now come to market is the proprietary cross laminated timber board product. Several brands are available with independently certified approvals, and these are starting to be used in bridge decks, as we have already seen in *Figure 4.13*, and as demonstrated in the Pont de Crest (*Figure 6.10* and Chapter 9). Consequently, these are likely quite often to supplant the more labour-intensive generic cross laminated deck in larger projects.

Figure 6.5 A: a three-layer cross laminated deck being built in situ, using sap-free sweet chestnut boards – Pont d'Ajoux. B: an area of five-layer cross laminated softwood deck similar to the type shown in BS EN 1995-2. In this case the individual boards have been pressure treated prior to assembly. Furthermore, to ensure greater continuity, structural finger joints may be included in individual boards so that each can be full-length
Photo A: © J. Anglade
Drawing B: © CJM/TRADA Technology

6.6 Nail laminated decks

Transverse nailed laminated decks (*Figure 6.6*), are another well-established type. They are satisfactory for relative light volumes of vehicular traffic, as well as for pedestrian, cyclist and equestrian bridges. The arrangement of the laminates in the direction perpendicular to traffic flow is important, so that loosening of the fixings does not occur with time. A longitudinal nail laminated arrangement may at first appear convenient, since in this way stringers or secondary beams might be eliminated, but this method has been unsatisfactory and is no longer recommended.

Figure 6.6 Transverse nailed laminated decks. A: building a deck in situ, with pressure preservative treated Caribbean pine. Part of the kerb has also been fitted using long electro-plated screws – rear right edge of deck.
B: a completed deck, using pressure preservative treated radiata pine, with untreated sap-free rauli running boards attached using countersunk stainless steel hexagonal-head screws
Photos © CJM

Figure 6.7 A small stressed laminated deck for a footbridge. Sawn segments of preservative treated larch with staggered butt joints to form a bowed shape. The Town-type parapet handrail adds to the overall stiffness of the structure. Scotland Photo © CJM

Transverse nailed laminated decks (*Figure 6.6*), as well as similar screw-fixed decks that have recently been introduced remain popular as an easily built component for rural road bridges where traffic frequencies are relatively low. In practice, these decks contribute to the overall stiffness of the bridge, although this is difficult to prove by analytical modelling. Certainly, they can help to brace an under-deck braced girder system. The laminations are nailed to one another with a skew pattern, forming a simple plate. To attach the deck to the main structure, nails or large screws are used. Running boards should be attached with countersunk stainless steel fixings or electro-coated recessed or octagonal-head screws – it is essential that replacement should be easy.

6.7 Stressed laminated deck design concepts

These are highly orthotropic components. For instance, with a stressed laminated deck made from sawn European softwood, the ratio E_{90}/E_0 is only 0.015, according to a table in Eurocode 5-2. Crews provides similar data based on full-sized tests using Australian radiata pine sawn laminates. Rather than having to perform a full orthotropic plate analysis, however, the engineer has the option of using a greatly simplified beam analogy method. The deck is modelled as a series of beams whose effective width and transformed mechanical properties take into account the transverse stiffness and shear resistance of the corresponding plate. To validate such procedures, full-sized ultimate load tests have been made in the structures laboratory, as well as in situ load-deflection measurements on real bridges.

For an assumed distributed width, located at the centre of the plate depth, point loads are taken down into the plate from the wheel distributions (for example, half an axle load). The assumed distributed width is further modified by dimensional constants reflecting the exact material type and manufacturing method. For example, nailed laminated decks, various stressed laminated types and timber–concrete composite decks all have different factors. Once this transformation process is complete, the deck, now effectively a set of beams, may be analysed conventionally according to its support details and any continuity present.

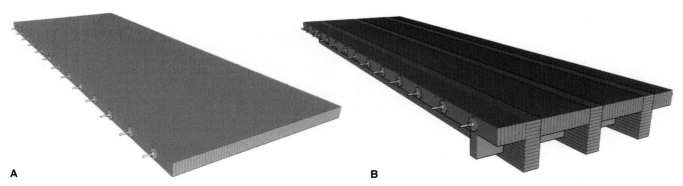

A **B**

Figure 6.8 Stressed laminated decks. A: a simple design comprising pre-stressed edgewise-arranged preservative treated boards – lengthways these may be finger jointed or butt jointed using rules to stagger the intervals – see Chapter 7.
B: a T-beam design incorporating flat-wise and depth-wise glulam elements, both types incorporated within the calculated clamping system
Drawing © CJM/TRADA Technology

For lightly loaded bridges and pedestrian walkways on larger structures, nailed or screwed crossed-layered plank decks are a useful alternative to the full-stressed laminated deck, since they do not require pre-stressing components. Two-layered arrangements are common, and a design method is included in Eurocode 5-2. A typical symmetrical crossing angle is 60°. For these decks, the intersection angle, the actual contact width of the point load, together with the distance between longitudinal support centres, are used in a simple expression leading to an effective width.

Figure 6.9 Stressed laminated decks have been developed in Australia using both indigenous hardwoods and radiata pine LVL
Photo © K.Crews

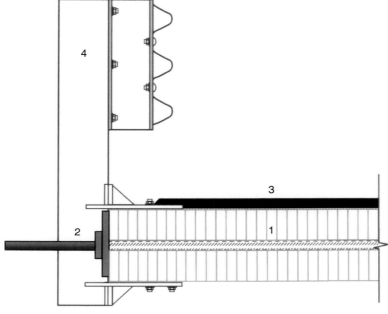

Figure 6.10 Part-section showing typical details of a stressed laminated road bridge deck
Key:
1 Sawn softwood or hardwood laminations; if softwood, approximately three hardwood edge strips are typical.
2 High-strength corrosion-protected pre-stressing bar with proprietary clamping system.
3 Approved geotextile membrane, edge seals and polymerised bitumen wearing surface.
4 Proprietary steel barrier items; steel or timber parapet posts.
Drawing © CJM/TRADA Technology

Figure 6.11 Part-section showing typical details of a cellular stressed laminated road bridge deck
Key:
1 Sawn softwood or hardwood laminations as in Figure 6.10.
2 Preservative-treated LVL webs, typically 65 mm thick, at 500 mm max. Centres. Diaphragms also of LVL at intervals along span.
Drawing © CJM/TRADA Technology

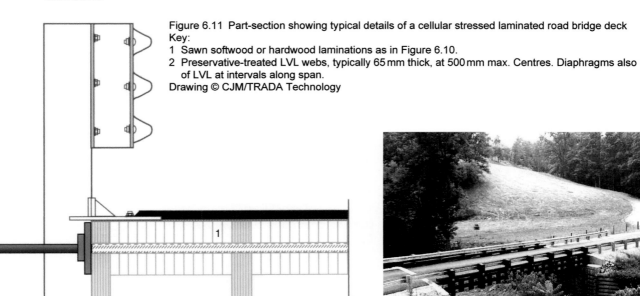

Figure 6.12 A hollow cellular stressed laminated deck providing a road bridge in West Virginia, USA. The solid timber laminations comprise North American red oak treated through an oil-based pressure preservative process
Photo © CJM

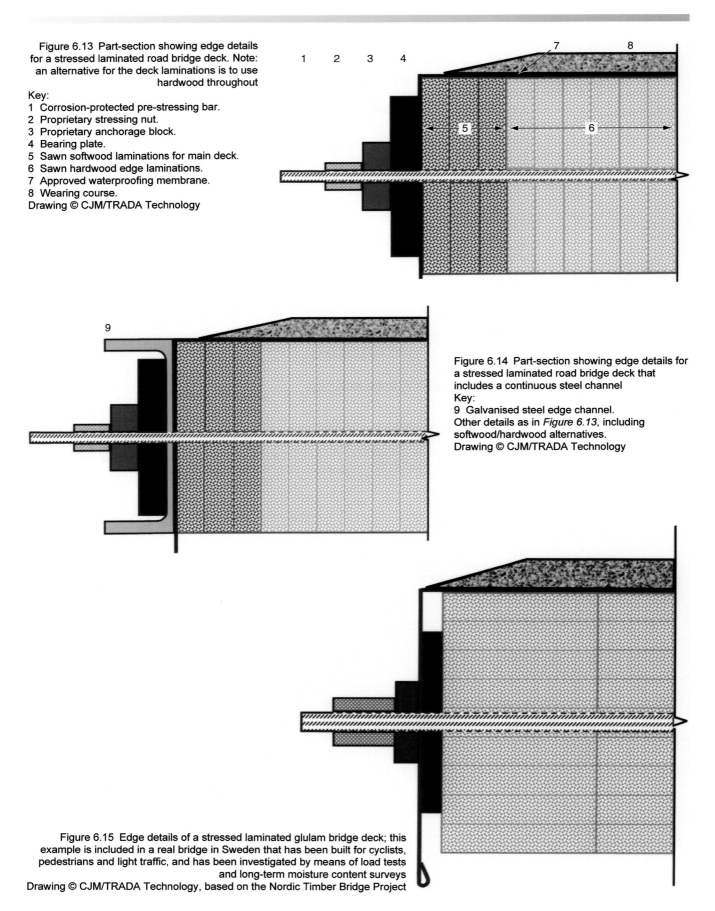

Figure 6.13 Part-section showing edge details for a stressed laminated road bridge deck. Note: an alternative for the deck laminations is to use hardwood throughout

Key:
1 Corrosion-protected pre-stressing bar.
2 Proprietary stressing nut.
3 Proprietary anchorage block.
4 Bearing plate.
5 Sawn softwood laminations for main deck.
6 Sawn hardwood edge laminations.
7 Approved waterproofing membrane.
8 Wearing course.
Drawing © CJM/TRADA Technology

Figure 6.14 Part-section showing edge details for a stressed laminated road bridge deck that includes a continuous steel channel
Key:
9 Galvanised steel edge channel.
Other details as in *Figure 6.13*, including softwood/hardwood alternatives.
Drawing © CJM/TRADA Technology

Figure 6.15 Edge details of a stressed laminated glulam bridge deck; this example is included in a real bridge in Sweden that has been built for cyclists, pedestrians and light traffic, and has been investigated by means of load tests and long-term moisture content surveys
Drawing © CJM/TRADA Technology, based on the Nordic Timber Bridge Project

Figure 6.16 A: close-up details of the underside of a typical modern European stressed laminated deck: showing the transverse steel tendons and clamping plates; the longitudinal glued laminations with staggered finger joints; the kiln-dried oak edge strips to improve evenness of pressure and offer high resistance perpendicular to grain. At the bottom of the photo is a simple non-stressed area of deck forming the footpath; Andelfingen Bridge, Switzerland.
B: an alternative way of obtaining high lateral compression on a stressed laminated deck – here, vertically laminated sections of beech plywood are used as edge strips – Leimbach Bridge, near Zurich, Switzerland
Photos © CJM

6.8 Stressed laminated glulam decks

Also included in Eurocode 5-2 are stress laminated glulam decks. These are similar to stress laminated solid timber types, but the sawn laminations are replaced with glulam slabs (effectively beams placed vertically, or with larger sections, on their sides). Because glulam provides higher design values and can be manufactured in larger sizes and lengths, longer spans are possible. Such decks are generally effective for spans up to about 18 m. The bridge described in *Figure 6.7* contains 215 × 360 × 16 200 mm laminated sections. The pre-stressing bars are size M24, strength class 8.8, and the initial lateral pre-stress on the glulam is 1.0 N/mm².

Stressed laminated decks may need to be re-tightened during routine maintenance. This topic has been subjected to applied research (*Figure 6.18*). In Europe, the technique adopted has been to install the laminates (either solid timber or glulam) at a moisture content lower than the mean ambient conditions. The subsequent swelling of the wood due to hygroscopicity then counteracts the creep losses in clamping force. This makes it possible to re-tighten only within the first year after installation, then achieving a reasonable period of 4–6 years before further re-tightening becomes necessary during routine maintenance.

Figure 6.17 A: prefabrication and the use of uniformly dry glulam elements is the approach adopted in Sweden, where this bi-curved stressed laminated deck was first assembled in the factory of Martinsons;
B: for erection at Umeå, a growing city in the north-east, on the Gulf of Bothnia
Photos © A. Lawrence

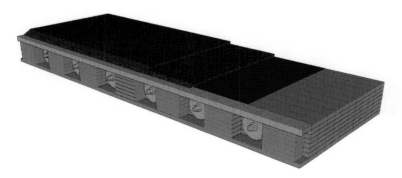

Figure 6.18 A sealed and waterproofed stress laminated glulam deck designed for use with a prefabricated parapet that is demountable for inspection and re-stressing; see *Figure 6.27* for further explanation
Drawing © CJM/TRADA Technology

Figure 6.19 The Vihantasalmi Bridge, 182 m long, carrying Highway 5 over a strait 180 km north of Helsinki. This has a composite timber–steel–concrete deck; the principal longerons are pressure preservative treated glulam; the vehicle deck is a reinforced concrete slab with special shear connectors into the longitudinal glulam beams; the transverse truss structures have timber lower chords with steel uppers and struts, to support the laminated boarded footpath. Finland
Photo © CJM

Another technique to ease the burden of re-stressing is the use of the demountable parapet assembly – see *Figures 6.18* and *6.27*. Stressed laminated decks may take the form of single decks with extended 'T-webs', and as cellular integrated decks using two compressed board layers joined by webs of LVL.

6.9 Timber–concrete composite decks

Timber with structural concrete, forming a composite deck, has been used for decades in, for example, New Zealand and North America. Now interest is growing elsewhere, including the Nordic region, where applied research, full-sized loading tests on actual bridges, and mature developed applications exist.

Early systems comprised nailed laminated decking with unreinforced concrete and a thin asphalt surface. More recently, thicker reinforced concrete layers including shear connectors have been added, giving greater composite action. Design rules for this form of construction are given in Eurocode 5 Part 1-1, the general design document, while supplementary rules, for example shear connection methods, are contained within BS EN 1995-2, *Bridges*.

As an alternative to designing and manufacturing for full composite action between the timberwork and the concrete wearing surface, the latter may be keyed onto the requisite timber elements using corrosion-resistant metal shear keys at suitably calculated intervals. These are often bonded into the timber using approved epoxy-family adhesives and hardeners, an approach taken, for instance, in the Pont de Merle, Chapter 3. *Figure 6.21* shows another example. This is an under-strutted frame bridge, with glulam longerons, transoms and struts and a prefabricated concrete slab deck keyed onto the beams.

Figure 6.20 This modern Norwegian trussed bridge with a timber–concrete composite deck has been designed for exceptionally heavy vehicle loadings, including tracked armoured fighting vehicles.
Photo © A. Lawrence

Figure 6.21 An under-strutted frame bridge, with glulam longerons, transoms and struts and a concrete slab deck keyed onto the beams. Switzerland
Photo © JPM TRADA Technology

Hodsdon and Walker investigated a low-cost form of composite deck using round timber and limecrete at the University of Bath in England. The project included prototype tests on full-scale bridge decks of a size suitable for small rural footbridges. Indicatively successful results were obtained, but the researchers made recommendations for further investigations that would be necessary before bringing the technology into wider use.

6.10 Innovative all-timber decks

As we saw in Chapter 4, a number of innovative materials and techniques are being introduced into decks. These include multi-layer board diaphragms, which are bonded and press laminated from sap-free softwoods (*Figure 6.22*).

Another innovation, the block laminated deck, has already been mentioned and is illustrated in *Figure 6.1*, while the case studies of Almere Pylon Bridge and Tharandt Botanical Gardens walkway in Chapter 9 also include them.

Their special advantages include the ability to provide three-dimensional curvature, through the double laminating process, and an exceptionally shallow height of section in side profile due to their structural efficiency.

6.11 Wearing surfaces for vehicular bridges

As a wearing surface for vehicular traffic, non-surfaced timber provides insufficient friction, particularly when wet or frozen. In addition to providing structural advantages, timber–concrete composite decks may be designated simply to obtain a suitable wearing surface.

As part of the Nordic Timber Bridge Project, a study was made of wearing surfaces of sealed timber decks. Both asphalt and concrete paving types were investigated. Stability against swelling or shrinkage in use is a concern, and this is achieved by making the deck plates from control-dried materials, including bonded laminates and, if necessary, applying re-stressing maintenance (essential when lateral tendons are included in a stressed laminated design).

Blistering of seals during application, due to water vapour rising from a timber substrate that is at too high a moisture content, is a problem that has been experienced occasionally. However, as indicated above, the deck structure should in any case be built with a moisture content not in excess of approximately 15%, for reasons of subsequent stability in service.

6.11.1 Sealing

As discussed in Chapter 3, the protection strategy starts by considering whether to permit water freely to run through the deck, or whether to seal it. For open-drainage designs, the arrangement needs to be well ventilated, slightly tilted and with distinct open gaps between the boarding. The risk evaluation process described in Chapter 3 should be followed, investigating whether to stipulate 'durable' or 'very durable' timber species, from which sapwood has been eliminated, or to use deck boards that have been pressure-preservative-treated with a suitable formulation and preservative loading.

Figure 6.22 A light traffic bridge using braced glulam longerons surmounted by a multi-layer board press laminated deck diaphragm using sap-free Douglas fir, sourced from the same region as the bridge location in which it is used – Crest, Drôme, France
Photo © CJM

Figure 6.23 An asphalt wearing surface above a membrane and a multi-layer board diaphragm; this deck also has a lateral inclination of 2.5%, to ensure drainage – Crest, France
Photo © CJM

A sealed timber or timber composite deck is normally protected and sealed using several layers. Generally these entail a membrane above the timber plate, an under-layer of waterproof material such as a polymer-modified bitumen composition, and a wearing surface for the traffic.

Where binding layers and asphalt pavements are employed, the Swedish Roads Administration authority has developed standard test procedures. These assess sealing layer adherence and through-thickness shear. Welded insulation mats and waterproofing layers have also been tested, to assess the effects of timber treatment on adherence, and to improve the surface finishes of the wood layers that receive the seals.

The use of reinforced concrete in conjunction with a timber deck plate has been described above. These types of deck also require the provision of good-quality sealing materials, with the correct specifications and workmanship.

Similar comments apply when pre-cast reinforced concrete slabs are laid over timber systems of timber stringers and transoms. In these cases, sealing and topping methods generally use flexible membranes, screeds and edge seals in accordance with normal concrete technology.

In a good-quality timber deck design, having provided a slightly tilted main surface to ensure lateral drainage, a longitudinal gutter is then required, with facilities to collect and drain-down the resulting run off at the ends.

6.12 Parapets and handrails

The following considerations affect the choice of footbridge parapets:

- the type of footbridge user – for example, pedestrians only, or equestrians/cyclists also;
- the nature of the site and locality – for example, whether it is a rural or urban location, and whether it passes over a main road, railway or a stream;
- rgulations and bridge user guidance – for example,concerning apertures.

In the case of bridges carrying vehicular traffic, there are further considerations, including more strict loading requirements for kerbs, parapets and handrails, taking into account impacts from collision.

Where footbridges cross highways, motorways or railways, there is often serious concern over objects being accidentally kicked, or deliberately dropped, onto the traffic lanes or rails. In applications such as these, authorities will stipulate the required dimensions of enclosure. These will normally prevent an open solution, but it is a simple matter to include light metal screening – see, for example, *Figure 6.25* and *Figure 6.26*. Alternatively, the parapets of some of the historic bridges in both timber and latticed ironwork have been glazed in shock-resistant safety glass, to preserve the open appearance while conforming with this demand. In rural areas and on cross-country footpaths, quite open bridge parapets are often acceptable.

Figure 6.24 A pylon bridge at Hochstetten, Germany, showing the recommended parapet height of 1400 mm required for a dual-purpose cyclist and pedestrian bridge
Photo © F. Miebach/Schaffitzel

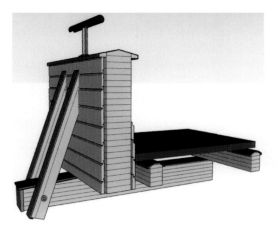

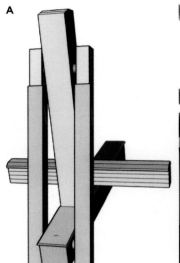

Figure 6.25 A braced parapet and hardwood handrail on a glulam trough-bridge. Note the protection arrangements and deck drainage
Drawing © CJM/TRADA Technology

6.12.1 Cantilevered parapets

Following the principle of 'horses for courses', the designer should consider the option of using steel elements for certain parts of the parapet. The obvious advantage of the material here is its high strength and stiffness, with a consequent lightening of the overall aesthetics of the bridge. Because of its mechanical properties, a cantilevered parapet is more likely to be structurally achievable, while plenty of efficient methods have been developed to attain strong connections between structural steel and timberwork. *Figure 6.29A* shows thin electro-galvanised steel parapet posts that have been moment-connected to a block laminated deck for a pedestrian and cyclist bridge. In *Figure 6.29B* is shown a similar parapet that was recently provided in the Netherlands. These types eliminate the necessity for diagonal timber stiffening braces, and can include stainless steel mesh, for example, to provide as much protection for traffic passing beneath as any local regulations may require.

Figure 6.26 Two types of braced parapet. A: twin struts and a single transom; for adequate lateral resistance, the struts are attached to the glulam post and the transom using calculated connections, e.g. dowel-type fasteners with shear plates – note also the spacer rings to ventilate the adjacent surfaces and the metal protective covers. B: struts in-plane with the post and sandwiched between paired transoms – note protective details of spacers between members, stood-off strap brackets and metal cap on post; also the metal flashing on deck edge, screen on tie beam sides and angle cuts on member ends – Essing Bridge
Drawing A: © CJM/TRADA Technology
Photo B: © A. Lawrence

Figure 6.27 A: here the immediate parapet post struts are paired, likewise the transom struts. Sandwiched between is a canted brace running down to the side face of the main glulams, providing a rigid triangle with perfect protection above.
B: above all this, a well-made round-section, pressure preservative treated handrail. Beneath the handrail is a slot passing to the centre of the section, to relieve stresses and thus avoid splitting; also to form the attachments for the metal screening. Both: Pont de Crest, France
Photos © CJM

Figure 6.28 A double-transom braced parapet structure in conjunction with a stressed laminated deck
Drawing © CJM/TRADA Technology

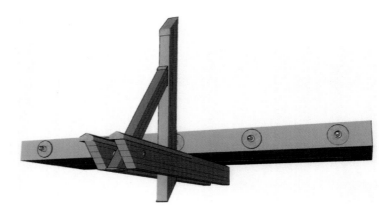

Figure 6.29 A: steel parapet posts, moment-connected to a block laminated deck for a pedestrian and cyclist bridge. Note also the protection on the edge of the deck.
B: a similar parapet recently provided in the Netherlands.
Photos © F. Miebach/Schaffitzel

6.12.2 Demountable parapets

These are a clever innovation to facilitate maintenance that has been developed within the Nordic Timber Bridge Project. The parapets are prefabricated (*Figure 6.30*), with posts that are connected, including air-gaps, to side attachment beams that match the depth of the main deck plate. Using hot-dipped galvanised steel plates inset to the attachment beams, a strong moment-resisting connection is achieved with bolting, eliminating the necessity for diagonal braces and allowing the units to be easily detached for inspection and, if necessary, re-tensioning of the deck units. Units similar to these have been used on several bridges carrying pedestrians and cyclists over railways in Sweden.

6.12.3 Handrails

Both softwoods and hardwoods are used for these (*Figure 6.31*), but the latter type, in a suitable species, is often preferred for durability and smoothness to touch. Most, if not all, of the smoothest-to-hand timbers used in joinery are of tropical origin. Recently, there has been good experience in the use of alder, a temperate and smooth timber of good durability, but not always easy to obtain. External weathering tends to aggravate splinter pick-up. For types of bridge that cater for mixed light motorised traffic, cyclists and pedestrians, a smooth banister-type handrail may be expected. *Figure 6.32* shows a parapet that includes part of the main structure, with the top horizontal of the railing frames proving sufficient handhold on a golf course bridge.

A

B

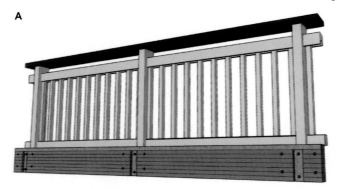

Figure 6.30 A: a partial bridge section showing a demountable timber parapet attached to a sealed and waterproofed stress laminated glulam deck.
B: a section of the prefabricated parapet viewed from the rear, showing how it is attached/detached for maintenance. Note various protective design features – the top edges of the attachment beams are covered; all the glulam is pressure preservative treated; the design includes stood-off weather-boarding outside the deck edges; and the softwood parapet parts, also pressure treated, are finished with a water-repellent stain. The hardwood handrail has a sloped upper surface, and a drip moulding has been milled in its lower face
Drawing © CJM/TRADA Technology

6.13 Ramps, steps and stairs

Ramps pose a significant aesthetic challenge, although structurally, the simplest option is often to frame them using the same techniques as the main crossing. The need for universal access, and the common occurrence of crossing requirements associated with flat sites, particularly relating to highways and railways, often makes them essential. It may be possible for ramps or steps to be accommodated within pathways that follow the contours of the adjacent topography, providing direct routes spanned by the structure. Manipulating the topography is probably the ideal way of minimising the impact of ramps. In this respect, taking care over route alignments at the early planning stage needs to be borne in mind.

Ramps should ideally be no steeper than a slope of 1:20, while 1:15 is considered an absolute upper limit of steepness. If the location means that greater gradients are unavoidable, continuous handrails, steps and landings may be considered. The disabled, visually impaired and elderly should be borne in mind. Risks arising directly or indirectly from slipping and losing foothold are significant.

Structural ramps should normally have a simple geometry, ideally along the desire line. It is usually preferable to maintain the same structural section for the ramps as for the main spans, even if this means a reduced structural efficiency. Broadly the same structural form, or one that reads in context with the main span, is also sensible. Curvature in the plan of the ramps, and indeed in the entire deck and pathway, can be another successful device to overcome gradient problems, especially on a restricted site.

Figure 6.31 A: a handrail turned from alder, with a parapet using stainless steel posts and cables, Norway;
B: a handrail of polished Douglas fir, Traunreut, Germany
Photo A: © CJM
Photo B: © F. Miebach/Schaffitzel

Figure 6.32 A parapet that includes part of the main structure – this arrangement is possible with through-trusses and girders. As shown here, the upper chord may directly support the handrail, but to achieve this the structural design has taken 'U-action' into account, so that lateral bracing above the deck is avoided. This ekki bridge carries pedestrians and light traffic for golf course maintenance and emergency access. Denham, Middlesex, England
Photo © CJM

Figure 6.33 A pedestrian and cycle crossing recently completed at Hoppegarten Station, on the S-Bahn system, centred on Berlin. Illustrating the inclusion of stairs and lifts in a timber and steel bridge. The basis of the design is for a combined deck of the materials to be suspended from an overhead glulam trough structure
Photos © F. Miebach/Schaffitzel

Figure 6.34 A: a roof incorporating discrete steel portal frames as well as timberwork, offering lateral wind resistance and bracing to the seemingly suspended glulam tied-arches. This roof in the French Alps has an interesting and somewhat oriental stepped appearance. St. Etienne en Dévoluy.
B: an erection view of a covered bridge at Ainring, Germany – the method of bracing is clear. Placing it well above traffic in this manner avoids risk of partial collapse if an accidental impact occurs
Photo A: © J. Anglade
Photo B: © F. Miebach/Schaffitzel

6.14 Roofs

In general terms, the roofs of fully covered bridges follow the broad principles for those on other buildings. However, some of the shapes and coverings that are especially useful for bridges are shown in *Figure 6.34*. *Figure 6.16* provides indications of the necessary shelter angles at the eaves.

As well as providing user protection, the roof may benefit the structural design of the entire bridge in terms of overall stability, robustness and arch stabilisation, for instance. This is demonstrated in both pictures in *Figure 6.34*, where the vital bracing effect is clear. This aspect is also seen in the illustrated examples for Passerelle de Vaires-sur-Marne (*Figures 5.6* and *5.7*) and the Punt la Resgia (Chapter 9).

7 Structural design

7.1 General

7.1.1 The partial factor method

BS EN 1990 defines the principles and main requirements for the safety, serviceability and durability of all types of structure. It is based on the limit state concept, used in conjunction with a particular partial factor method. Internationally, other non-identical partial factor codes exist. Eurocodes BS EN 1991–1999 inclusive conform to the Eurocode 0 basis. These developments have taken place within the framework of the European Committee for Standardisation (CEN), and it is an important and progressive step for the BS EN 1995 set of codes for timber structures to adopt an identical basis of design to that for the other main materials.

7.1.2 Bridges

The design of timber bridges should follow the Eurocodes and their associated standards (called 'Normative References'). The basis for the design of these has been improved by the publication of BS EN 1995-2, *Timber Bridges*, where previously no national standard existed specifically for this material, and BS 5400 was often adapted, with various difficulties. In addition to the specific Part 2, BS EN 1995-1-1, *General: Common Rules and Rules for Buildings*, is an important reference for the timber bridge designer.

This is because Part 1-1 addresses common items such as flexural, tensile and compressive members, connections and bracing, following a Eurocodes principle that similar information is not repeated in different parts of the codes. Hence Part 2 deals only with items that are especially intended for bridges – such as stress laminated deck plates – noting that otherwise 'unless specifically stated [the general part] applies'.

As indicated in Chapters 5 and 6, decks need not necessarily comprise just simple plank-boarded types, since crossed laminated and stress laminated versions are widely used and are included in BS EN 1995-2. Another useful deck alternative is the timber–concrete composite type, also addressed in the Eurocode 5 Part 2. Innovations continue – for example, the bi-curved block laminated deck shown in *Figure 7.1* has already been in service in Germany and adjacent countries for several decades and is proposed for inclusion in a revised edition of BS EN 14080.

Figure 7.1 General view of a curved prefabricated block laminated deck ready to leave the factory. Not yet directly referenced by the Eurocodes, components such as this are accepted by type approvals, linked to the *Construction Products Directive*. They are generally designed using the principles of BS EN 1990

Photo © F. Miebach/Schaffitzel

Having determined a conceptual solution (Chapter 5), consultations with potential suppliers and manufacturers should be extended. At this stage, the team should further consider materials preferences and availability; exact standardised members sizes (which vary from one manufacturer to another, in the case of glulam for example); and the strength classes that are readily available, all while continuing to be concerned with aspects of durability. Preliminary design, which – as opposed to 'concept design' – is a stage considered to belong to this chapter, should also entail the development of method statements for fabrication and erection.

For all types of timber structure, care needs to be taken over designing and fabricating the connections because both the design office time and the manufacturing work are time-consuming and should only be done once! An efficient and economical choice at the initial design stage is a significant factor in overall success. For convenience, both preliminary and full design aspects of timber bridge connections are grouped below in Section 7.7, but it is expected that all aspects will be appreciated before the project team embarks on a fully committed design.

Detailed engineering design follows preliminary design, and this includes comprehensive structural analysis. This work should only commence once there is full agreement, particularly by the client or his representative. This is because it is relatively easy for changes to be made during the concept and preliminary phases, but they become very expensive once the engineering office is committed. Production information such as the development of full manufacturing and erection drawings or their electronic equivalents is accompanied by work to take all of the quantities, linked to formal specifications that can be checked during workshop inspections and deliveries.

7.2 Preliminary design sequence

In conjunction with previous plans and surveys, now to be brought closer to completion, the team will further confirm the functional requirements of the bridge. This will establish the shape of the main structure – not only whether it is to be a beam or flat girder structure, or a suspended form, arch or truss, but also the level of the deck in relation to the crossing datum and the main structure. The plan of the deck – whether right, curved or skewed – will also be determined. The relationships between the members of different components should be examined – how they support one another and how they cross, bearing in mind functionality, connection arrangements and durability. Other aspects include whether there is a need for ramped or stepped approaches and what are to be the clearances and set-backs. User access requirements will be fully clarified and protection rules defined for all of the bridge users, e.g. parapet heights and enclosures.

Following these decisions, the general arrangement drawings will be produced, during which a very important consideration is how overall stability is to be achieved. This is further addressed in Section 7.3. Through this general route, further definitions are set out for the main outlines of the structure: length; span(s); deck widths; number of lanes and/or traffic separations, e.g. pedestrians, cyclists, equestrians; emergency traffic dimensions and access. Using loading codes such as the BS EN 1991 suite and its National Annexes, the engineers will classify the loadings – for timber, this will entail at an early stage an awareness of the associated implicit duration of critical load cases.

Table 7.1 Preliminary design sequence

No.	Sequential steps
1	Location – in conjunction with site maps/plans and ground/water profiles – possibly piers/abutments/harbour/dock walls to be re-used – fully establish the functional requirements of the bridge, including shapes – right, curved or skewed plan; flat, ramped or stepped approaches; bowed or flat deck; clearances and set-backs
2	With overall elevations and plans from Step 1, define dimensions to provide main outlines of the structure: length; span(s); number of lanes and/or traffic separations, e.g. pedestrians, cyclists, equestrians; emergency traffic dimensions and access
3	Classify loadings, access requirements, protection rules for/from all bridge users, e.g. parapet heights and enclosures
4	Select candidate structural systems to suit above parameters, making approximate estimates for shapes and sectional sizes of principal structural members. Include principles of stabilisation – see Section 7.3
5	Using 'Risk Analysis' approach – see Chapter 5 – investigate alternatives for materials, protection strategy, robustness of design against accidental or deliberate local damage, full-scale impact risks, redundancy and reserves of strength for long-term safety
6	Prepare preliminary method statements for fabrication and erection

Table 7.1 is an outline summary of this preliminary design sequence. Moving towards the position of seeking full design approval, the team needs to obtain agreement over the choices of materials with respect to potential durability, costs and maintenance requirements. The robustness of design against accidental or deliberate local damage, full-scale impact risks, redundancy and reserves of strength for long-term safety will also be considered.

7.3 Principles of stabilisation

For all bridges, overall stability is vital – as long, slender structures they are particularly prone to weakness and unserviceability if they are deficient. As we saw in Chapter 2, various historic devices such as arch-stiffened trusses were used, and in Chapter 5 for modern bridges, when considering the salient outline design features, chord stability and wind resistance remain the primary concerns. Uplift, sliding and overturning effects produced by the wind must be resisted, both during execution and for the finished structure. The main structural assemblies in the bridge have major load-carrying duties, making the stiffness and strength of the compressive and flexural beams and chords an essential challenge.

The general rules for structural timber design in BS EN 1995-1-1, which do of course apply equally to bridges as well as buildings, invoke three general principles:

1 Structures which are not otherwise adequately stiff shall be braced to prevent instability or excessive deflection.
2 The stresses caused by geometrical and structural imperfections, and by induced deflections (including any joint slip) shall be taken into account.
3 The bracing forces shall be determined on the basis of the most unfavourable combination of structural imperfections and induced deflections.

With regard to point 1, certain types of bridge can be shown to have structural stability through their inherent stiffness without the necessity for separate triangulated wind and stability bracing. For points 2 and 3, the 'general rules' code addresses both single members in compression and the bracing of beam or truss systems. Each is covered in Clause 9.2.5 Bracing. Also, for the latter, while a timber bridge may comprise only one pair of main beam, truss or arch frames, these can be treated by the same rules, thus considering them as a 'system'.

For the basis of these aspects of design, a simply supported, initially straight column with intermediate elastic supports was considered. This is a classic problem in the theories of elastic stability. Certain practical assumptions, such as the introduction into the expressions of the design axial load-carrying capacity rather than the theoretical Euler load, are described in the background paper by Larsen.

In principle, the code rules require the engineer to compute the strength of the supports (i.e. the design stabilising force) for single members in compression and for the compressive edges of rectangular beams, and to do this it is necessary to know the maximum initial deviation from straightness between supports. Fortunately, the recommended limits of $a/500$ for glulam or LVL members, and $a/300$ for other types are within the manufacturing limits provided by normal, quality-assured producers. Consequently, it is often unnecessary for the engineer to give this matter special attention, but it does merit awareness. At the time of writing, a large set of EN standards pertaining to glulam are being amalgamated into a new version of the Harmonised Standard, BS EN 14080, giving further information on permitted deviations.

7.3.1 Bracing forces

For General Rule 3, a constant in the code expression for minimum spring stiffness at each intermediate support is taken to be a nationally determined parameter (NDP). Introducing the value indicated in the UK National Annex leads to:

$$C = (4 \, N_d)/a$$

where C is this minimum spring stiffness, N_d is the mean design compressive force in the element, and a is the bay length.

As well as requiring certain stiffness, the design also needs to take into account a minimum stabilising force. Summarising this for glulam compression chords and girders (the reader may consult the code and its relevant National Annex for other types), and inserting another NDP, this amounts to:

$$F_d = (N_d)/100$$

where F_d is the required design stabilising force and N_d is as above.

For the compressive edge of a rectangular-sectioned beam, a comparable stabilising force may be computed. Requiring a knowledge of the maximum design moment acting on the member, this introduces the lateral buckling factor $k_{crit,}$ for which rules are given in BS EN 1995-1-1 Clause 6.3.3(4), including the possibility of assigning unity if the compressive edge is adequately restrained and provided torsion is prevented at the supports.

For beam or truss systems, in bridges often including pairs, the code assumes that there will be lateral supports at intermediate nodes. An expression is provided for the required resistance to both external horizontal loads (e.g. wind), and the internal stability load per unit length. A limit is given to the allowed horizontal deflection of the bracing system. To complete these expressions, a set of NDPs found in National Annexes is required, and for the United Kingdom, these are within the general range stated in the body of the code.

7.3.2 Means of providing stability

Having set out these principles, typically, how are they satisfied? In essence there are three routes towards ensuring stability:

1 triangulation
2 portal principles – with moment fixity
3 the use of diaphragms.

In the Norwegian military-loading truss bridge illustrated in *Figure 7.2*, the solutions are relatively straightforward, because the structure enjoys a high under-clearance, permitting the trusses to be located below a timber–concrete composite deck that acts as a diaphragm. The firm stabilisation of the compressive chords is thus readily secured, while the triangulated lateral wind bracing can easily be connected to the support posts and bottom chords.

Loads from the deck diaphragm are transmitted to the ground through a triangulation system, working with the support columns. Wherever the general site profiles permit, a system of this type should always be first choice, since as well as these other conveniences, the timber trusses are well protected from the elements by the deck above. Note that this includes a significant 'rain shadow' or overhang area. Note also that these glulams are double-treated (for further information, see Flisa Bridge in Chapter 9).

Figure 7.2 This military-loading truss bridge in Norway illustrates the advantage of an under-deck timber structure, to be considered whenever clearances permit – the stability of the compressive chords is easily secured since they are not remote from the deck; the lateral wind bracing, connected to the support posts and bottom chords, is readily included without interrupting traffic, and the trusses are well protected from the elements. Of course, flood clearance for the bottom chords is essential! Another advantage of this system is that as many parallel trussed frames as are necessary for the total loading can be added without complicating the general arrangement
Photo © A. Lawrence

Figure 7.3 For this simple open-decked glulam beam bridge of 14 m span and 2 m effective deck width, lateral stability for the main beams and the horizontal wind resistance of the bridge is provided by the triangulated wind girder system below the deck. Effective lengths between nodes are taken conventionally, bracing strength and stiffness is calculated using BS EN 1995-1-1 Section 9.2.5, while Section 6.3.3 provides calculation methods for bi-axial bending and torsional stability
Drawing © CJM/TRADA Technology

Figure 7.4 Where the functional geometry – clearances, etc. – permits, overhead bracing may be possible, but as illustrated here, a design solution is required for headroom at the lower-level entry and exit positions above the deck. In this case, the unbraced final segments of the upper chords are restrained by direct connection to sets of stiffened steel end-portals that are clamped to the foundations
Drawing © CJM/TRADA Technology

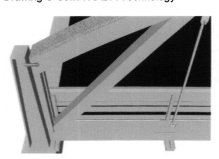

7.3.3 Triangulation

Often, in smaller beam bridges, open-boarded decks are used, so no diaphragm exists. An example of this is shown in *Figure 7.3*. Here, lateral stability is ensured for the main glulam beams, which are protected with chamfered top surfaces because a species that is only moderately durable (e.g. larch) is used. The horizontal wind resistance of the bridge is provided by the triangulated wind girder system located below the deck. Because of its protection, this girder structure will not be damaged by rainwater percolating through the decking gaps. Unlike the glulam beams, the deck boarding is a 'very durable' hardwood type, containing castellated non-slip profiles with inserted mineral strips. These boards have been mounted on elastomeric strips to eliminate local vibrations and to ensure firm, tight locking-down of the deck screws. Correctly fitted, this deck type is capable of resisting full-service class 3 exposure. If deemed necessary, the vibrational response of this type of bridge can easily be checked using BS EN 1995-2 Annex B, although for a footbridge of this size and proportions, it is unlikely to be critical.

We have already met in Chapter 5 the type of overhead braced structure illustrated in *Figure 7.4*. As illustrated here, a design solution is required for headroom at the lower-level entry and exit positions of these types. The unbraced final segments of the upper chords are restrained by direct connection to sets of stiffened steel end-portals that are considered to be clamped to the foundations. The general view of Flisa Road Bridge (*Figure 9.12*) illustrates another example of such a steel portal bracing system. This can be seen in the photograph, at the entry to the foreground truss, just above the intermediate pier.

These general principles apply equally to overhead braced arches – see, for example, the 70 m main span Tynset Bridge shown in *Figure 2.24*. However, in this case, necessitated even more by the considerable size and load-carrying capacity, the arches are also stiffened laterally in their lower segments, by increasing their breadth and designing the bearing pins accordingly.

7.3.4 Portal principles

The bridge type shown in *Figure 7.4* effectively combines triangulation – in the overhead zones where clearance is not a problem – with portal principles to solve the problem of the descending bottom chord segments, as explained above. The key issue is the stiffness of the portals. Another implementation of portal principles applies in what are sometimes known as 'trough bridges'. These usually entail deep, solid, glulam beams that form part of the parapet as well as providing the spanning structure, but bridges employing 'U-frame' action are also formed with horizontal girders, often the modified Warren type (*Figure 7.5*). Lateral torsional stiffness is achieved with a series of, in this case ten, moment-stiff portals. For the rigid connections between the transoms and the truss verticals, inserted

steel flitch plates are involved (*Figure 7.5B*). The fasteners are plain (but corrosion-protected) steel dowels, 16 mm diameter in this case. Standard timber engineering methods of designing moment connections are followed. These involve the fasteners rotating about a centroid in resisting the imposed lever arms, and not forgetting to provide for direct forces – in this case the transoms also transmit vertical shear from the deck loading.

7.3.5 Diaphragms
Figure 7.2 has already shown a bridge that relies on diaphragm action. Another, whose part-section is shown in *Figure 7.6*, achieves lateral stability entirely through diaphragm action. The deck acts as a two-way spanning plate. This is one of the stress laminated deck plates illustrated in BS EN 1995-2 Clause 1.5.2.3, type d, with pre-stressed glued laminated beams positioned edgewise.

The design principles and application rules for stress laminated deck plates are given in Clause 6.1.2 of Part 2 of Eurocode 5. As a matter of principle, the long-term pre-stressing forces are such that no inter-laminar slip occurs, hence the EI_{z-z} can be fully mobilised for any necessary calculations involving resistance to horizontal wind loading. Since there is plate action, then clearly lateral torsional instability does not arise, once it is established that the supports are correctly designed.

Note that this example also includes a detachable prefabricated moment-resistant parapet (see Chapter 6). This type has been developed in Sweden. During routine inspection, the posts and rails are relatively easily removed as units, allowing if necessary the re-tensioning of the pre-stressing tendons.

7.4 Analysing the main structural systems

7.4.1 Preliminary design
BS EN 1995-1-1 Section 5, *Basis of Structural Analysis*, contains quite sophisticated guidance that should satisfy the most thorough checking. However, all but the largest of timber bridge designs, such as those shown in the Nordic region in Chapters 2 and 9, normally begin life as simply supported beam models and pin-jointed structural frames whose in-plane arrangements are essentially stable and structurally determinate. For these types, very approximate shape, span range and member proportions are readily available to initiate the analysis. For example, the *Manual for the Design of Timber Structures to Eurocode 5* published by the Institution of Structural Engineers and TRADA contains charts for trusses, pitched-tied

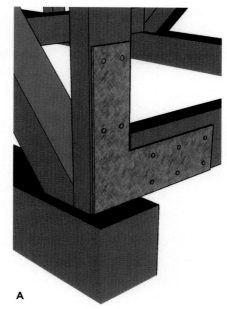

Figure 7.5 A: achieving lateral torsional stiffness of the compression chords of a modified Warren truss by means of 'U-frame' action – effectively a series of, in this case ten, moment-stiff portals; B: for the moment connections between the transoms and the truss verticals, inserted steel flitch plates contain plain (but corrosion-protected) lateral steel dowels, typically in the order of 16 mm diamater. In this diagram, the front portal is sectioned laterally to reveal the inserted plates. Omitted for clarity is a triangulated horizontal 'wind truss' between the bottom chords, although if a diaphragm-type deck were to be used, this might be omitted
Drawing © CJM/TRADA Technology
Photo © CJM

Figure 7.6 A partial section (with foreground parapet assembly omitted) of a stress laminated deck bridge formed from a large series of straight horizontally glued laminated elements – in this example the typical deck thickness is 320 mm. Restraint stability and lateral wind resistances are both ensured through two-way spanning plate action designed in accordance with BS EN 1995-2
Drawing © CJM/TRADA Technology

frames, flat girders and arches. Individual glulam manufacturers and associations also provide such charts. Many of the illustrations in this publication mention main sizes and give further clues to proportions.

During preliminary design, potential control by serviceability limit states should not be overlooked since timber structures are relatively light and flexible. Roughly checking the deformations of the members and components using unfactored loads and mean stiffness values will often estimate the viability of a scheme. These types of checks will also indicate whether, at a later stage, second-order analytical models as described in the code will be required. For ultimate limit states control, an approximate indication can often be achieved if not obvious by inspection, through a technique familiar to timber engineers. Every load case has to be assigned a specific duration of load classification. Consequently, the modification factor k_{mod} can be introduced at an early stage – for bridges, often assigning it a short-term duration. Then the action effects of each trial load case can be divided by the associated k_{mod} to determine whether it is critical. When doing this, the shortest-term action involved in the combination for the load case is always that to be tested.

7.4.2 Detail design

The *Manual for the Design of Timber Structures to Eurocode 5* gives guidance on how to derive characteristic values of actions, design values of actions and actions through the Eurocodes suite. For bridges, some additional references will be made to Annexes of BS EN 1990 and 1991, to access the appropriate action combination factors (ψ values).

As we saw in Section 7.3, some parts of the frames may involve moment-stiff connections, and the final checking of these may necessitate analyses involving linear or rotational connection slip, including long-term checks. These functions will formally satisfy the principles of the code stated in sub-clauses 5.1 (1)P and (4)P. Even here, however, the words 'commensurate with . . . the reliability of the information on which the design is based' are included. So, as stated many times above, it is advisable before entering a refined analysis, to produce preliminary designs. These may remain somewhat approximate, but should be reasonably close to a final solution, so that the design team and the client's representative, as well as potential manufacturers, can agree everything. The design manual referenced above contains a discussion of computer frame analysis techniques (Section 3.2.4), including choices of stiffness values, elastic constants and spring stiffness methods using codified slip factors.

7.4.3 Form-specific analysis aspects

In many respects, the design of timber beam-columns using Eurocode 5 is quite similar to the general approach for steel structures. This includes the principles of achieving stability, described above. A discussion of effective lengths for bending members together with nomograms for k_{crit} may be found in *Manual for the Design of Timber Structures to Eurocode 5*.

For arches, BS EN 1995-1-1 contains guidance on assumed initial deviations in geometry – see its figure 5.3 and its associated sub-clause. With trussed forms, individual element lengths are often shorter than with beam or slab and diaphragm bridges, and even the principal top and bottom chords of the truss are often sub-divided at certain nodes. For analysis, triangulated frames are often treated as flat planes with intersecting 'system lines' (see general code Section 5.4.2, *Frame Structures*).

7.4.4 Preliminary calculations

In association with an initial analysis, in which the action effects are determined in the normal manner, using classic structural mechanics, the influence of cross-sectional slenderness and compressive-flexural stability should be taken into account. Here, the recommendations given in BS EN 1995-1-1 *General* are appropriate. These methods are based on decades of experience, originally thoroughly researched and backed with full-sized structural tests and computer simulations. Simple expressions are available that may be used for preliminary calculations and later refined for confirmation of the design.

Engineering manuals for use with BS EN 1995-1-1 include charts showing buckling modification factors plotted against slenderness ratio, which may be used for rapid estimates. Members subject to axial compression and bending about both axes can be validated in this way. Such procedures are likely to be necessary for bridge arches when these are subjected to combinations arising from vertical loads plus wind. Guidance on the analysis of plane frames and arches also includes methods and recommended parameters for initial deviations in geometry and the effects of induced deflections on internal forces.

7.4.5 Stability of arches

For the initial appraisal of resistance to buckling in the vertical plane, an effective length of approximately 1.1 times the half-arc length is recommended for a two-hinged arch, and 1.2 times this value for a three-hinged structure. For horizontal buckling, the principles described in Section 7.3 are followed. The lateral bracing configuration, together with the fixity at the bases (in plan), are influential factors. Using truss-braced cross struts between the arches, the length between the bracing attachment nodes can be used for checks at higher levels on the arches or between tall trusses (e.g. Beston Bridge). These are the positions where bracing structures can be attached without traffic interference. At positions closer to the deck, the dimension from the lowest bracing node to the arch or frame footing is taken, for use with alternative buckling length ratios.

Two methods of laterally stiffening arches are common. Where a substantial deck-width-to-arch-span ratio exists, the lower arch segments are broadened. This is the method used in the Tynset arches, for example (*Figure 2.24*) – this decision also entailed special design work for the main support bearings, to ensure resistance to lateral twisting. Footbridges are often slender in plan-to-span ratio (*Figure 7.7*), so individual arched frames may be canted towards one another. The resulting alteration in the geometry in itself makes them stiffer, because some of the lateral action effects are now resisted in two-way bending. Furthermore, the arch profiles may be increased towards the bases, but rather than increasing their breadth, as in the first method (which increases manufacturing costs), it is preferable to increase their depth. By controlling the outlines in these ways, an elegant structure can be created, while at the same time ensuring acceptable lateral stiffness.

Figure 7.7 Canting arched frames together to improve lateral stability and wind resistance. Moment-fixed steel tubes, with end-flanges, are attached to inner side faces – Millennium Footbridge, Pervou, Switzerland
Photo © CJM

7.5 Continuing the analysis: validation stage

Before the full calculations are started, approval in principle will have been granted, and a series of checkpoints will have been answered to ensure that the preliminary scheme design is feasible, and is likely to lead to a safe, serviceable and durable outcome. Practices will have their own internal procedures, but the following is a typical set of criteria that should have been answered before validation can commence:

- Have the construction logistics been satisfied?
- Have the workshop/factory manufacturing and prefabrication phases, delivery and erection demands been planned, bearing in mind site access, ground conditions, lifting and temporary propping equipment, weather risks, etc.?
- Do the preliminary manufacturing, delivery and erection costs meet requirements?
- Have the site assembly nodes been identified, relating these to the erection methods statements?
- Have 'sustainability' and chain of custody materials certification requirements been considered?

As indicated above, if validation proceeds before these questions are satisfied, there is a serious risk of time and effort being wasted. Provided the answers to checklists such as this are satisfactory, the engineers will proceed according to *Table 7.1*, and *Figure 7.8*, which outline the steps to produce more complete timber bridge design calculations using BS EN 1995-1-1 and its Normative References. Connection design is just as important as sizing the members, because a great variety of types are possible in timber, only some of which are suitable in bridges, and much time will be lost on false starts unless these are addressed firmly.

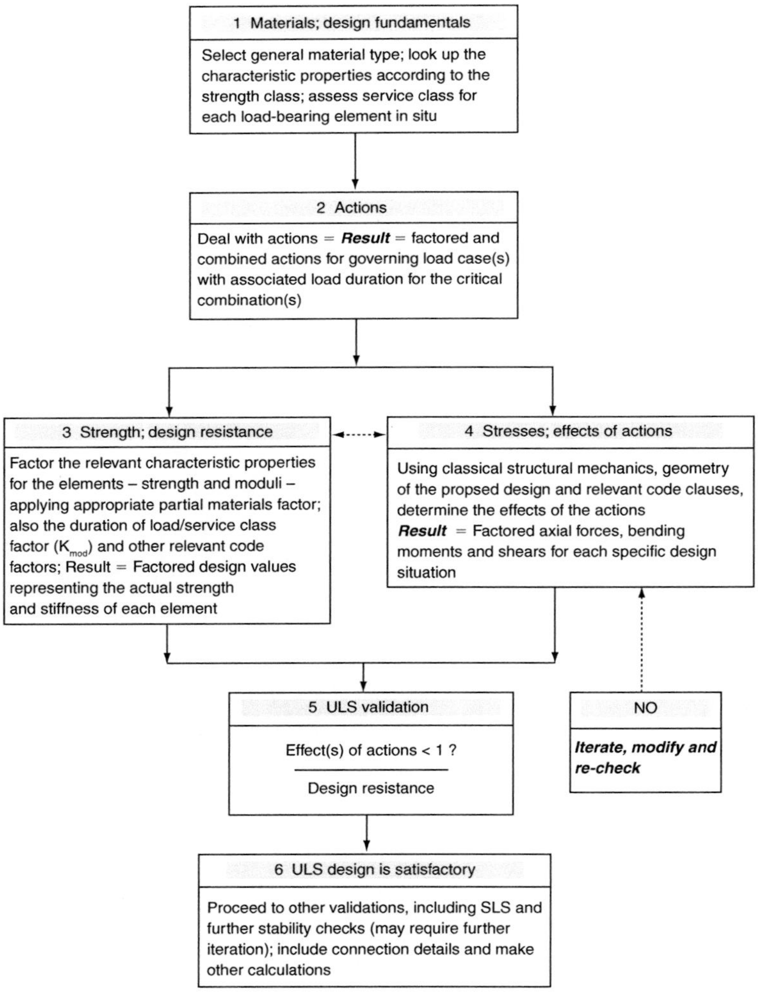

The engineers should obtain agreement from all parties to the proposed scheme and check all information leading to and coming from the steps in *Table 7.2*. They will then start to prepare lists of design data, specifications and drawing/database notes. Proceeding to connection design; special deck design (BS EN 1995 Part 2 – e.g. stressed laminated, timber–concrete); and further detailed design work – e.g. extra ULS and serviceability-limits-states checks, bearings at supports, notches, tension perpendicular (curved glulam), diaphragm actions, fire resistance, if required. Details will be provided for the deck drainage and sealing measures – see Chapters 5 and 6, but noting that these also affect secondary member spans and deflections. See, for example, *Figure 7.4*, where the deck board spans have been halved by including a longitudinal runner of the same 'very durable' hardwood as the deck boards themselves.

7.5.1 FEM calculations

Finite element modelling is a valuable analysis tool provided it is not used as a complete substitute for experienced care. It can help to

Figure 7.8 General procedure for the validation of an approved scheme design using Eurocodes

provide solutions for some of the more advanced design issues, including estimated lateral instability forces when plates are present, and also for obtaining indications of the natural frequencies. Haiman describes the calculations and finite element computer simulations provided for a 17 m span, 5.90 m wide road bridge that was recently built near Zagreb. This structure has a pre-stressed glulam deck, four main glulam carriage beams, and a protected, sealed running surface. The FEM model had orthotropic elastic parameters representative of timber (an important point to ensure in these cases), and the pre-stressing of the deck was simulated. The accelerated gravitational load, from the total mass of the bridge, yielded natural frequency predictions. Strength, stability and static

Table 7.2 Design calculations checklist for design validation using BS EN 1995-1-1 and its Normative References

1 Ensure bridge structural form is fundamentally stable under all potential actions. Identify all load-bearing members – have precautions been taken against decay? See Chapter 3	For forms, see Chapters 2 and 5. Durability precautions affect engineering design – species/material type; self-weights (increased by preservative content); strength classes. Obtain characteristic properties from appropriate BS EN, according to the general material type (see Chapter 4)
2 Calculate the distribution of the variable actions to each structural component and member	Analysis is likely to be iterative. An approximate analysis will have been made for preliminary design (*Table 7.1*). At the validation stage, the analysis and its consequences will be fully developed in stages that depend on the complexity of the structure. Stage 4 will result in fully factored load cases, but these may initially be for ULS only. SLS checks (e.g. see Stage 10 below) require differently adjusted load sets
3 Determine service class of elements and components	The service class for a bridge element or component depends on its precise location. Not all parts of a timber bridge are necessarily in the same service class as one another. Protective design detailing usually aims to place parts in service class 2, rather than 3
4 Deal with actions (loads) – develop factored and combined action combinations, with associated load duration for each critical case	The structure's purpose type influences the way that the action combinations are computed – the combination factors (ψ values) differ for a footbridge compared with various types of building. Before the actions are combined, each load has an individually associated duration. Then, the 'load duration for the critical combination' is determined by the shortest-term of the individual actions within the load case. See other design guides for further advice on dealing with actions
5 Ensure all potentially critical action combinations are considered	Several combinations are likely to be investigated, and it may be advisable to check special ULS situations, e.g. accidental actions (impact for example); stability cases; and fundamental ULS conditions (e.g. flexural moment resistance). Initially it may not be clear which case is 'critical' – validation of several conditions may be necessary before this is ascertained
6 Determine partial materials factors – National Annex of BS EN 1995-2	Partial materials factors (γ_M values) vary according to the general material type.
7 Determine the duration of load factor for each key element	The factor k_{mod} that adjusts strengths for load duration actually depends on the material type, the load duration for the critical action combination (see Stages 3 and 4), and the service class for the element (Stage 2)
8 Determine other relevant design factors for each key element and cross-sectional axis (y–y; z–z; both combined)	These may include, for example, k_h (depth/width), k_{crit} (for slender beams or those not fully restrained at ends), k_c (compression members, arches), k_{sys} (load sharing or system strength), etc.
9 Compare the factored design values representing the actual strength and stiffness of each element with the factored axial forces, bending moments and shears	For each specific design situation – including, e.g. resistance to buckling about either or both transverse axes (compression members, arches); resistance to lateral torsional instability (beams, decks); resistance to torsion couples (parapets, bracing, bridge roof if present)
10 Modify and re-check, as necessary	Complete other validations, including SLS checks – static and dynamic (vibrations – modes, damping ratios) and further overall stability checks; connection details and connection calculations

deflection checks were also supported. As we shall see in Section 7.6, pedestrian 'comfort criteria' (against excessive vibrations) are regarded as satisfied if natural frequencies are above 5 Hz (vertical). This was the case for this example, with first, second and third mode values being 7.2 Hz, 8.6 Hz and 17.4 Hz, respectively.

7.6 Fulfilling serviceability conditions

As indicated above, since timber structures are relatively light and flexible, approximate serviceability design checks should begin at the preliminary design stage. According to the general principles of BS EN 1990, unfactored loads and mean stiffness values are used for these purposes. Also, mean values of materials' density are taken. For dynamic serviceability calculations, the variable actions from pedestrians and other traffic are normally omitted, and only the full mass of the structure is included. For the final calculations, this may take into account superstructure items such as kerbs and parapets, plus sealing under-layers, e.g. diaphragms of structural plywood or LVL, asphalt layers, etc., all of which may helpfully increase mass and add to the damping effect.

7.6.1 Limiting values for deflections

BS EN 1995-2 Section 7 *Serviceability Limit States* contains a short sub-clause on these limits, permitting national choice and referring to National Annexes. For UK designs, these have been determined in the British Standard National Annex and are summarised here in *Table 7.3*. These are said to be recommended limits *'for deflections due to traffic load only'*.

In allowing for creep in timber structures, the combination factor Ψ_2 forms part of the calculations for the quasi-permanent (long-term) action combination and for bridges, appropriate values of factors Ψ are given in the National Annex to BS EN 1990:2002/A1:2005. For cycle track loads, and all other types of traffic load, wind loads and snow loads, this parameter is set to zero, clarifying the fact that these effects are omitted from the long-term serviceability calculations. But creep calculations using the bridge mass are advisable.

For these, the irreversible final deflection can be calculated in the normal manner for timber, using the k_{def} values indicated in BS EN 1995-1-1 table 3.2 and its accompanying text. For bridges, unfortunately, no published recommended limits are available for the final deflections. With open-decked bridges intended mainly or exclusively for pedestrian use, the engineer might consider applying the suggested final limit of l/250 as recommended for the more strict situations stated in the building applications National Annex. With a sealed deck and a bridge intended for more frequent motorised traffic, a more strict long-term limit of perhaps l/400 might be chosen.

Table 7.3 Limiting values for deflections (National Annex)

Actions	Limiting value for deflections in beams or trusses of span *l*
Characteristic traffic load	l/400
Pedestrian load and low traffic load	l/200

7.6.2 Limiting vibrations caused by pedestrians

BS EN 1995-2 Section 7 *Serviceability Limit States* has a brief sub-clause on this subject, making reference to a simplified method for assessing such vibrations, which is contained in Annex B of this code. Simple references to national choice are of no consequence in this case because the UK National Annex accepts both the simplified method and the recommended damping ratios (*'where no other values have been verified'* – meaning designers are at liberty to have tests done).

The timber bridges code cross-refers to BS EN 1990:2002/A1 for consideration of 'comfort criteria'. This simply means that vertical or horizontal vibrations should not unduly alarm bridge users, which is of course a very subjective matter. In general terms, and for normal cases, provided that the checks described in the simplified method are carried out, then the criteria expressed in the Eurocode 0 documents, in terms of acceleration maxima for individuals or persons in crowds, are unlikely to be exceeded.

Vertical resonance is most likely to occur when the frequency approximates to the human step frequency, or double this value. Horizontal resonance should be suspected with a natural frequency around the human pace, or half of this value. Runners and joggers are also provided for in the BS EN 1995-2 Annex B method, which was developed from research and practical measurements on real timber bridges at the University of Munich.

As indicated in this reference, the simplest way of ensuring that pedestrian 'comfort criteria' are satisfied is to check whether natural frequencies are above 5 Hz (vertical) and 2.5 Hz (horizontal and torsional effects). For a simple open-decked beam footbridge, such as that shown in *Figure 7.4*, a very approximate indication may obtained using a 'beam analogy'. The conventional formulae of engineering physics may be used. Experience suggests that for such types, the deck and other 'semi-structural items' contribute about a further 40% to the bare EI of the set of longitudinal beams. More precise estimates for frequency may of course be made using structural modelling software, and for larger or more bespoke designs, particularly those with diaphragm decks, this is the route likely to be adopted.

7.6.3 Large pedestrian bridges, and vibrations experienced by pedestrians on vehicular bridges

Special dynamic investigations, beyond the scope of this advice, but following similar principles with regard to human response, will clearly be necessary in the design of very large pedestrian bridges – that is, those in the order of 40 m or more span. Likewise, special forms such as the pedestrian and cycle tension-ribbon bridge, Main-Donau Canal, Essing, Germany (73 m maximum span) will require extensive investigations into all aspects of serviceability performance, including the avoidance of disturbing vibrations and the influence of wind effects. Generally it is the case that vibrations experienced by pedestrians on vehicular bridges are not perceived as unpleasant because of the stiffness necessary for large axle load. Nevertheless, this also warrants some consideration at the detailed engineering stage.

A

Figure 7.9 Typical truss node connections in Norwegian light traffic road bridges. Steel suspension ties are used in both of these designs (A: Beston Bridge detail; B: Nesoddveien Bridge, near Oslo). The slots are for relatively thin steel plates – typically 8 mm each, and the dowels are typically 12 mm. Electrolytic galvanising and epoxy coatings are applied, while the slots are open-bottomed to facilitate drainage. The structures were to a large extent prefabricated. Note also the substantial protection of the upper edges of the double-treated glulams
Photos © CJM

7.7 Connection design

Compared with steelwork, numerous options exist, and fabricating firms are generally smaller and more diverse, each having their own preferred techniques and specialist equipment. Consequently, consultation on connections should begin during the earliest phases of the project. When alternative schemes are being assessed, architecturally, for functional suitability and for durability, the connection options should be included straight away, rather than being treated as a detail to be settled later. Furthermore, as a highly engineered structure, the buildability and overall costs of the bridge are strongly influenced by the choice of connections.

7.7.1 Connection design fundamentals

Compared with the theoretical 'thin-wire, pinned node' models of elementary structural mechanics, the timber engineer has the task of connecting members with real lateral dimensions. A finite height of section, or width on plan, necessitates the examination of overlap areas or arranging the abutting of members against one another, using techniques that facilitate this choice. Nowadays, timber bridges and other large frames in this material tend not to be fabricated using simple side-lapped joints and fasteners such as plain bolts and ring connectors. With these, there are several disadvantages, and these have led to improvements that now place elements in a more generally straight arrangement on plan. Nails, bolts and the older types of side-lapping connector often have too low a load capacity in wide-spanning and heavy-duty applications. They also often give rise to significant slip under load, and generally form shapes that have eccentricities or offset moments in plan.

Preferred contemporary timber bridge connections often involve the use of relatively thin steel plates that are inserted into slots sawn into the connected members during factory controlled processes. Jigs are used to guide the slitting saws, and plain steel dowels are fitted through holes in both the steel plates and the timber. It is possible to have numerous parallel plates within each connection, giving a high load-carrying capacity and stiffness. Sometimes, when an internal member in a truss or arch carries

Figure 7.10 Multiple flitched-in plates secured with dowels are used on large Norwegian bridges, earlier introduced in the 1994 Winter Olympics stadia to carry forces of up to 7500 kN. This type of arrangement enables all of the elements to intersect in a plane – a considerable advantage over shear plate and split ring connectors. Most modern timber bridge manufacturers use similar techniques, and these are covered by Eurocode 5. Special cutting equipment is available, and designers need to be aware of typical details as well as the calculation basis. In the arrangements adopted for the Evenstad Bridge, clearance slots pass right through the lower chords of the trusses, enabling the tension hangers to be connected to steel transoms
Photos © CJM

Figure 7.11 A: jig-guided saw to cut connection slots in glulam, Oregon, USA; B: a base connection using flitched-in plates and plain dowels for a compression member in a glulam framed bridge at Wuelflingen, Switzerland; note the effective drainage and an elastomeric bearing plate
Photos © CJM

pure axial forces, the manufacturer may add one or a few bolts with round washers and nuts, instead of the plain dowel locations. This technique is used if a set of slots in the member is open at one end, as a precaution against fatigue loosening. It is not normally done in a chord member, where slots are usually enclosed at both ends by the surrounding timber.

7.7.2 True pins

True pins are necessary at most heavily loaded nodes (*Figure 7.12*), and these are achieved using steel fabrications similar to those found in all-metal structures, although for timber, the actual sizes of the plates, forgings and tubes tend to be smaller due to the lower structural mass. Methods of attachment include the use of timber connectors such as single-sided shear plates, flitched-in arrangements of various formats, and where a service class 2 condition can be established, e.g. under a roof or sheltering deck, then bonded-in threaded rod attachments are useful.

Where a service class 2 condition can be established, other connection methods used in bridges include large finger joints and bonded-in rods. BS EN 1995-1-1, which specifies BS EN 387 as the standard giving their production requirements, references large finger joints, but restricts their application to service class 2 situations. These types of finger joint enable complete glulam or LVL elements to be end-jointed, unlike the smaller types of finger joint used in individual laminations. As indicated in the code, service class 2 conditions lead to an average moisture content in most softwoods not in excess of 20%. In a bridge, this can be achieved with

Figure 7.12 A: true pinned cable support under block laminated deck – Jagstbrücke, Möckmühl, Germany;
B: pinned end details connected via inserted plates and steel pins, Warren truss covered bridge, Traunreut, Germany;
C: pins and steel plates connected under the timber deck using single-sided shear plates, Hochstetten
Photos © F. Miebach/Schaffitzel

Figure 7.13 Innovative materials and techniques in an uncovered arched road bridge – San Nicla, Tschlin, Switzerland, crossing the river Inn. For each transom, beech plywood is spliced to the remainder of the softwood glulam – the former gives great compression strength perpendicular to the grain. The protected and threaded steel rods connect the deck hangers to the arches. The whole assembly is sheltered from external exposure by a combination of larch cladding and metal covers
Photo © CJM

suitable protection, for instance from an overlying waterproofed deck, or from local or complete roof covering. However, the use of large structural finger joints should not be contemplated without significant consultation and careful evaluation of the manufacturer's experience.

An example in which large finger joints were used is in the tension-ribbon footbridge over the Main-Danube canal, near Essing in Germany. For this large structure, nine $220 \times 650\,mm$ section glulam tension-beams were spliced, under covers, close to the site, to a total length of up to $73\,m$.

In sheltered positions within the bridge, bonded-in rod connections can also be considered. In the Douglas fir glulam beam bridge at Crest, Drône, France, for example, these have been employed to connect the 'tree-like' timber struts into the reinforced concrete piers. Such connections can also be considered to attach steel pods or fins to the bases of members, so that these can be raised above splash zones where the timbers are connected to the upper levels of the foundations or pier caps. This was the technique used to connect the glulam arch bases in the San Nicla Bridge over the river Inn.

Design methods for bonded-in rods have been established through a thorough applied research project, and the technique originates from applications in wind turbine technology. Given suitable adhesive formulations, and the necessary manufacturing skills to ensure cleanliness and proof-testing, it is a technique that can now be employed with confidence.

Nailed or screwed connections in timber bridges are generally only appropriate for the connection of the smaller and less heavy-duty components. In the design of decks, partial or full roofs and protective screens, suitable structural wood-based panel products may be involved. As indicated in Chapter 4, LVL is a generic material recognised by BS EN 1995-1-1, and this may have a role in the design of decks, diaphragms and roofs, particularly since pressure preservative treatments have been developed. These extensions to the design scope mean that not only timber–timber, and timber–steel–timber connections can be involved, but also those involving solid timber and glulam with these further materials. Fortunately, the connection design methods given in the Eurocodes now operate with a harmonised theory that caters for all dowel-type mechanical fasteners and wood materials with a range and mixture of embedment strengths. These latter are developed from a knowledge of the strength class of the materials concerned, and their characteristic densities.

7.8 Fastener and connector design

7.8.1 Choice of fasteners and connectors
Details on metalware items are generally confined to steel types, which are those predominantly used. BS EN 1995-1-1 calls for compliance with two harmonised European Standards, namely BS EN 14592 (dowel-type fasteners), and BS EN 14545 (connectors).

The brief Section 4 *Durability* of the Eurocode section on materials contains a table that gives examples of minimum specifications for the corrosion protection of fasteners, related to ISO 2081. Within the broad categories provided, stainless steel types are included, and these are clearly appropriate in certain applications, particularly in service class 3 situations and those where it is intended to use acidic species of timber. Care in specification is needed, however, since according to the grade, stainless steel may be less strong and with poorer ductility.

BS EN 14592 contains ranges of dimensions, tolerances and other requirements for nails, screws, plain dowels and bolts with nuts. Procedures for evaluation of conformity will be given, along with marking and certification information. In BS EN 14545 will be found similar details for shear connectors – plate, ring and toothed plate forms; and also for punched metal plate fasteners and nailing plates.

Eurocode 5 provides calculation expressions for the characteristic values of the yield moment in combined bending and tension for common standard types of nail, bolt, plain dowels and screws. These expressions also involve application of partial factors for material safety and load duration, as well as calculations for the embedment strength of the timber or wood-based material itself. When unusual types of fastener are proposed, even if these are claimed to have high performance, enquiry should be made of the manufacturers as to their suitability with timber structures. When the yield strength of the fastener is greater than normal, brittleness may be an issue, since the theory used by the code assumes that the fastener deforms significantly before ultimately yielding plastically.

7.8.2 Corrosion protection

Apart from metals that are inherently corrosion-resistant, such as stainless steel, protective coating should always be stipulated for fasteners and connectors used in bridges (*Figure 7.14*). Alternatives include hot-dip galvanising, sherardising and electroplating with zinc and cadmium. For bolts and plain dowels, electroplating is the preferred method, since the other treatments make turning of the nuts difficult and the protective layers may be stripped in the process. Guidance is provided in BS EN 1995-1-1 Section 4.2 *Resistance to Corrosion*. BS EN 14592 *Fasteners* and BS 14545 *Connectors* also contain sections on treatment, and on suitable specifications for stainless steel devices.

Figure 7.14 Protection of the heavy-duty moment connections in the Breslau cantilevered bridge comprises both electro-galvanising and an epoxy-based paint system; note also that the steel plates are divided into four separate layers to take into account movements across the grain of the timber
Photo © J. Schmees

7.9 Validating connection design

For connections, the same fundamental principles apply as are used in element design. For example, it is necessary to determine the service class of the region of the structure under consideration, and to decide on the duration of load class associated with the governing action effects. In this manner, vital modification factors, including k_{mod}, are quantified. The partial materials safety factor, γ_M, also has to be applied in order to develop the design resistance of the connection. It is important not to overlook these factors, but BS EN 1995-1-1 does not strongly emphasise their importance since it is assumed that the engineer is familiar with its general procedures. The designer may get lost in details on mechanical connection design, found in Section 8, without realising that Sections 2 and 3, which designate these essential factors, are also applicable.

Figure 7.15 Top chord of a mechanically laminated ekki Warren truss bridge. In members carrying axial forces with open-ended slotted joints, some of the plain steel dowels are substituted by bolts, nuts and round washers, as a precaution against the connection peeling apart by withdrawal of the dowels; such failures are unknown, however, and this is probably very cautious
Photo © CJM

For preliminary design, the general approach is indicated in *Table 7.4*. The basis of design uses expressions that relate to a series of failure modes. Each relevant mode is checked and the design resistance is developed from the lowest value. For inserted steel plates and dowels of the proportions normal in bridges, three typical failure modes are common: embedment; development of a single plastic hinge in the fastener; or failure through triple plastic hinges. Normally the axial resistance of the fasteners is included within the expressions leading to the characteristic values, but according to BS EN 1995-2, this so-called 'rope effect' is not included for bridges. This is understood to be a precaution against fatigue failure, possibly exacerbated by concerns over effects of moisture changes, although no evidence exists for withdrawal failures ever having occurred.

Because of the variety of configurations and factors involved, it is impossible to reduce timber connections to an elementary level, and for the consequent time-consuming work involved, design software is especially useful. Aspects need to be included such as the angle of the resultant forces to the grain, the number of shear planes and the number of fasteners in a line, as well as the partial factors indicated above.

Specialist advice may well be necessary at the preliminary design stage, and the best course of action is to consult potential manufacturers. These have the necessary technical expertise and the dedicated software that can explore all of the optimal arrangements and bring into play all the necessary factors, to develop a safe and efficient design. Checks on resistance to splitting and on shear resistance parallel to the grain within the connection zone also need to be made in the final stages.

Table 7.4 Connections with laterally loaded dowel-type steel fasteners – design calculation outline

Step	Notes
1 Propose general connection arrangement	Categories: single shear, double shear, multiple shear; timber–timber, steel–timber–steel, timber–steel–timber, multiple combinations. Timber–steel–timber most favourable for heavy duty – see illustrations
2 Select fastener type, diameter, material composition*: large screws, plain steel dowels, bolts and timber engineering washers and nuts, bolt sets with ring connectors	The preference is for plain dowels in significant numbers and relatively small diameters.** BS EN 1995-2 *Bridges* forbids the use of staples and punched metal plate fasteners. Screws, bolts and timber connectors are useful for smaller bridge components and replaceable parts. Design rules for all of these types are given in BS EN 1995-2 *General*, with the bridges part of Eurocode 5 dealing only with special items, e.g. connectors for timber–concrete composite decks
3 Arrange estimated number of fasteners in a grid, so that spacing, end- and edge-distances can be satisfied. Determine arrangements in plan-section to find thickness of each layer	Design resistances depend in part on spacing dimensions and are only valid if the parameters designated by the code are followed. These vary according to connection type. For some types, e.g. plain dowels, spaces may be reduced with consequent lower resistance per fastener. 'Loaded' and 'unloaded' ends and edges should be observed
4 Determine characteristic resistance per fastener for choice in Step 3 and modify to develop design resistance for the set***	Factors include: strength class (gives characteristic density); yield strength of fasteners; load duration; partial materials safety factor; number of shear planes; number of fasteners in line; angle of resultant force to grain in zone of connection concerned, taking into account any necessary eccentricities

Notes

Iterate Steps 1–4 as necessary, and check final solution for availability, fit, fastener shapes, quality approvals, durability factors, protective coatings, etc.

* It is important that the material composition of the steel fastener should lead to a plastic failure mode, since this is the basis of the design theory. Higher-strength compositions may bring about brittle failures that can only be validated by special testing for the design in question.

** In a large node connection in the Flisa Bridge, there are six layers of 9 mm thick steel plates within inserted slots. These are electro-galvanised and spray-coated. Stainless steel dowels of plain shank form are of 12 mm diameter. The centre distance between steel plates within the members is 80 mm. This dimension is optimised by software that gives the best balance between fastener yield strength and timber embedment resistance.

*** For preliminary design estimates, the resistance at 0° to grain and at 90° to grain can be determined from design manuals and/or software. These can then be linearly interpolated for intermediate angles with little loss of efficiency. The 'number of fasteners in a line' factor is quite significant, but has to be determined by trial and error. A starting value of around 0.65 is suggested.

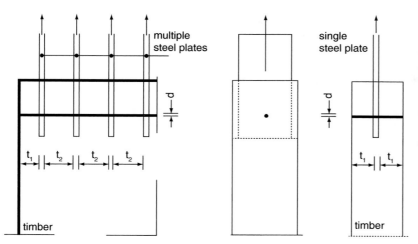

Figure 7.16 Connections with internal steel plates. Left: multiple steel plates; right: single steel plate. For the multiple type, the simple failure modes for the single type need to be adapted and checked step-by-step, working through the connection from side to side. In this manner, an optimum set of proportions is determined for maximum load-carrying capacity and stiffness – for example, the proportion of t_2 to plate thickness and dowel diameter d. Manufacturers have specialist software to carry out these functions; the general practitioner can obtain initial solutions based on BS EN 1995-1-1 Section 8, with assistance from design aids, e.g. from TRADA.
Drawing © T&F/TRADA, after Jorissen

7.10 Durability of connections

Durability detailing needs to be considered right at the start of preliminary structural design (*Figure 7.10*), because several alternative arrangements are commonly possible with regard to the shear planes and inter-faces. Some alternatives are potentially significantly better with respect to durability than others. In general, inserted steel plates are the preferred option, and these are configured in various ways, dependent on the structural duty.

Where there are connections between bridge elements and surfaces near the ground or the water, durability detailing is especially important. Common occurrences are where supporting struts, columns or arch bases rest upon pier or abutment caps, and other similar zones that provide the reactions to the timber structure from the upper foundation levels. These points need to be carefully detailed so that uptake of water into the

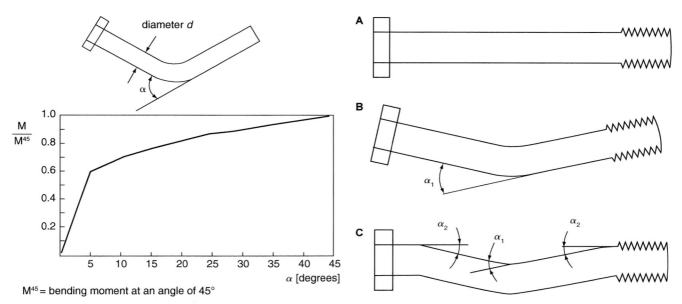

M^{45} = bending moment at an angle of 45°

Figure 7.17 Theoretical yielded shapes for a bolt – similar principles apply to plain dowels.
A: no hinge, embedment in timber;
B: single hinge;
C: triple hinge. For A and B, design limits are placed on angle α
Drawing © T&F/TRADA, after Jorissen

Figure 7.18 Good connection detailing for durability – Black Dog Halt SUSTRANS Bridge
Photo © CJM

vulnerable end-grain is avoided, and the bases are well ventilated. In the following guidelines, raising the lower ends of timber members at least 150 mm above the upper course of the supports is suggested, as an absolute minimum. But the actual height should vary according to the design and situation of the concrete, masonry or metalwork. For footbridges in particular, locations are often surrounded in summer months by lush grass and undergrowth, and such vegetation may encourage the growth of moulds and mosses near the bases of the timbers, with this in turn then leading to the risk of actual fungal attack. Lower ends of timbers supporting the bridge should be raised at least 150 mm above concrete or masonry surfaces, and these should be battened to ensure drainage away from the wood.

- Steel supports or shoes should be detailed to include ventilation and drainage slots, large enough, and well positioned so that they cannot become blocked by debris and dirt.
- Moisture-proof membranes or plates resting on elastomeric bearings may be useful; these can be attached to concrete or masonry that is drained as indicated above.

At higher levels in the timber superstructure, protective design of connections remains important. Slots in the timber to receive inserted steel connecting plates are left open on the undersides and are provided with a positive drainage incline behind the plates, where the member is tilted – for example, in girder diagonals. In any instance where a countersunk hole is provided for a bolt or dowel in a side-grain, the opening must either be plugged after the metal item is fitted, or alternatively it may be possible to cut a distinctly conical sink, so that there is drainage. To improve ventilation, posts connected with bolts or large screws may be 'stood-off' the parapet items to which they are attached, and for side-lapping beam-to-column connections, similar arrangements may be considered.

7.10.1 Moisture movement of timber around multi-dowelled plates

In designing steel plate connections for timberwork, a common error made by those used to steel structures is to make these plates too thick and too heavy. There is also a tendency to clamp the timber elements too rigidly, so that there is no capability of accommodating changes in section width and breadth through moisture movement. Typically, a 4% change in moisture content within a 900 mm deep glulam beam will entail approximately 20 mm alteration in this dimension – either swelling or shrinkage is possible, dependent on the conditions. When these natural adjustments are invoked by changes in ambient conditions, the considerable internal forces in the timber cannot be prevented by fixings and friction, so tight restraints simply lead to the material pulling itself apart across the relatively weak fibre direction. Such repeated movements due to diurnal and seasonal changes in temperature and humidity lead to growing cracks and fissures.

The solution is to include slots, rather than circularly drilled holes in certain locations within plates. In other instances, it may be possible to support a beam, for instance, simply from its lower edges, with lateral slots in the steelwork merely acting as locators to prevent misalignment in service. Another option useful for attachments of arch bases to their structural supports is to use lateral guidance in a similar manner to this, with the base of the arch member formed into a semi-circular shape so that it can roll under structural actions. A pair of steel straps may be used for arch crown connections and to form moment-stiff splices at erection nodes in chords. This gives free widths of timber between the metal parts, so that natural changes in the width of the wood are not tightly clamped.

7.11 The final details

The manufacturer will often undertake some of the final steps, once he has been identified and appointed. But care is important during the preparation of tender documents and the drafting of agreements, so that unfortunate disputes do not arise. Even though it may be more efficient to allow the manufacturers and suppliers to apply their specialist expertise, the division of responsibilities between them and the designers must be clear.

Production of final details includes checking and re-designing connections (if necessary), as well as detail-drawing them (or producing equivalent electronic versions); detailing and specifying bearings, load hangers, accessories such as protective metal covers, parapet and guide rail parts may also be undertaken by these specialists. However, this is generally at the instigation of the designers, who should have made sufficient allowances in the costs for the types of protective items that have featured heavily in this publication. Workshop/factory manufacturing notes, quantities and transfer of dimensional and design data for prefabrication are likely to be the responsibility of the engineering office.

8 Conservation, maintenance and repair

8.1 Importance of inspection and maintenance

The old adage that 'a stitch in time saves nine' is the key message of this chapter. All bridges, including those constructed with timber, should be inspected and maintained periodically. It is unfortunate that the benefits of these activities are often unappreciated. Also unfortunate is that when budgets are restricted, inspection and maintenance programmes are amongst the first to be curtailed. This only results in greater expense in the long run, as well as wastage of human, material and environmental resources.

Maintenance requirements are, in general terms, no different to those with other materials. Agencies responsible for public bridges generally operate material-independent procurement procedures incorporating the requirement for suppliers to formally document maintenance schedules. Private bridge owners are well advised to take note of this, and to ensure similar precautions.

Based on this publication and its references, it is possible to draft supplementary detailed inspection and maintenance specifications for timber structures. Inspectors of timber bridges need some training in the subject (*Figure 8.1*), although it is not difficult to learn the basics. Preventative maintenance, important for all bridges, means ensuring that moisture content levels of all timber and wood-based materials stay at a low level, as discussed in Chapter 3. It is important that the sheltering of key elements is maintained or even improved. Additional screening has been added to the formerly exposed side faces of the tie beams of the Main-Donau tension-ribbon bridge. More normal maintenance activities include ensuring that dirt and debris is regularly cleaned away, cladding and finishes are re-decorated and connections are kept tight.

Work indicated by inspections should be carried out regularly, to ensure that the structure remains in a condition that will give optimum performance and service. Effective inspection and maintenance ensures the public safety of bridges, extends their service life and reduces the likelihood of a sudden urgent programme of expensive repairs and remedial work. Hence significant savings in whole-life costs are achieved.

Figure 8.1 Modern techniques of timber inspection include the use of specialist instruments such as this decay detection drill. A full range of remedial measures exists
Photo © TRADA Technology

8.2 The influence of the structural form

As has been shown in the preceding chapters, the general arrangements and structural forms of timber bridges have advanced significantly over the past few decades. Awareness of the necessity of protection by design has been especially significant. Laminated decks, for example, have greatly improved the opportunity to create and maintain durable timber bridges, since these make it possible to seal the exposed upward-facing running surface. By strictly limiting deflections and building them with the correct membranes, sub-surfaces and wearing courses (as discussed in Chapter 6), seepage onto the structure below is prevented. The use of laminating materials installed at a low moisture content is another way of achieving this aim.

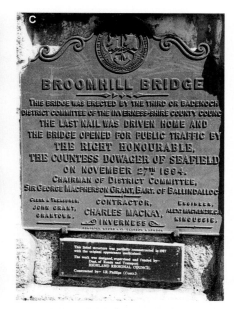

Good maintenance has its roots in the very early stages of a project, and Chapters 3, 5 and 6 have indicated how careful design will ensure a prolonged structural lifetime. A general consideration is to ensure that the proposed design, i.e. the principal form of the structure and the arrangement of the major bridge parts, is amenable to the type of maintenance that can reasonably be expected. Also, consider at the preliminary design stage how subsequent work can safely and easily be performed. See, for example, the prefabricated, detachable parapet assembly shown in Figure 6.3, which facilitates re-stressing of the deck and maintenance of the parapet itself. Effective maintenance stems too from the adoption of correct construction specifications. Another example is in the use right from the start of readily re-applied surface treatments that have been developed specifically for use on timber (Chapter 3).

8.3 Maintenance and construction specifications

At the tendering stage of a project, prospective contractors are often obliged by technical approving authorities to include information relating to inspection and maintenance procedures. This type of information may include:

- lists of specifications used in conjunction with the design – codes, technical standards and abstracts of their scope and assumptions;
- essential extracts from the materials standards stating what has actually been specified and used, e.g. timber or glulam strength classes; grading rules; dimensional standards; moisture content specifications;
- specifications for timber preservative treatments and the associated checking, control and certification measures;
- diagrams relating to intended pre-cambers or bows; planned setting-out and datum points;
- details, location and assembly procedures, showing the key connections – these may be required in future for partial or complete dismantling.

Figure 8.2 Broomhill Bridge, crossing the Spey in Inverness-shire, is a rare surviving example of a once-common kingpost truss type. This listed structure still carries a public road for motorised traffic. In 1987 certain trusses, now still in good condition, were replaced using frames connected by code-designed shear plates and greenheart timbers. This was not a 'like-for-like' repair, since the original parts of the structure are locally forested, creosote-treated Scots pine. But at least steel girders were not inserted!
A: general view of the bridge;
B: Scots pine piles at the supports, generally in good condition;
C: the commemorative plaques of 1894 and 1987;
D: the replacement trusses;
E: an example of the type of detail to be conserved whenever possible – in this case an original carpentry splice
Photos © CJM

Figure 8.3 Pyrmont Bridge, Darling Harbour, Australia, built between 1899 and 1902, and repaired to form part of the city monorail system. These approach spans in ironheart timber are said to represent the highest level of development of 'traditional' timber engineered trusses – this is probably true even taking into account early 20th-century designs in North America and Europe. The swing span is one of the largest in the world, and the first to be powered by electricity
Photos © CJM

8.4 Erection sequence: inspection and maintenance database

Submissions for larger projects are likely to be required to give a method statement for erection, which, if kept on file, contributes significantly when the time comes for inspection and maintenance. More major repairs may entail dismantling or moving parts of the structure (*Figure 8.3*), in which case it will be important to know how it was put together in the first place. A schedule of self-weights of the primary components is thus an important feature of such records.

8.5 Construction stage

Many future maintenance difficulties can be avoided simply by ensuring that the bridge is built to design and specification. Furthermore, as soon as the construction is complete, and before final payment, the client's agent should ensure that he is in possession of the type of information indicated in Section 8.3, and that the client is given a basic maintenance plan. This will vary in detail according to the scope and size of the project. However, it should at least indicate the details of expected regularity of inspection, and cover items such as the renewal of finishes and the intended design life of any bridge parts whose durability is deliberately set to a lower level than that of the major structure itself.

Measures should also be taken immediately before and during construction, which will help to ensure that a basis for good maintenance exists on record. These include the following:

- Ensuring that permitted variations to specifications (both with regard to materials and also concerning erection and assembly), are recorded, and that an accurate set of 'as-built' drawings are produced and filed. These may beneficially include small (e.g. A4) simplified plans and elevation drawings, which can form the basis for inspection instructions.
- Completing a bridge record sheet, which lists sensitive regions of the structure; items whose design life is deliberately intended to be less than that of the major structural parts; and expected regular checks, e.g. re-tightening of connections, re-assessment of pre-tensioning forces, checking and possibly replacing cladding.
- Compiling a set of photographs to record construction work, particularly that which is by its very nature covered up, so that any subsequent deterioration may be assessed by comparison with the original condition.

8.6 Inspection and preventative maintenance

As a matter of policy, authorities responsible for publicly accessible bridges generally specify inspection intervals. As already suggested, other types of owner should follow this example. On the 'stitch in time saves nine' principle, there should also be a policy to recognise and attend immediately to minor maintenance work that is indicated in between these main intervals. Troive suggests a typical interval of six years for all types of public access bridge. In situations where large numbers of travellers cross at frequent intervals, such as in the approach to station quays, more frequent inspection and maintenance may be appropriate.

8.6.1 Techniques

A range of techniques is available, varying from the use of simple wood testing instruments such as picks and moisture meters, through more sophisticated methods such as the decay detection drill shown in *Figure 8.1*. Crews describes how a dynamic-based testing method has been used to make reliable assessments of aged timber road bridges. This involves the attachment of accelerometers beneath the bridge structure, which are then excited by a modal hammer connected to portable instrumentation. From estimates thus obtained of the residual stiffness of the bridge, correlation techniques predict the remaining bending moment capacity with probabilistic relationships and reliability theory.

8.6.2 Recording

A standard 'bridge inspection form' normally includes details such as the location and age of the bridge, and its structural form, access and position of the deck. It has sections for noting the construction and condition of the substructure and superstructure and any decay of the timber members.

From a survey carried out in 2004 in Scandinavia, where hundreds of relatively new timber bridges have been installed, *Table 8.1* shows the most common issues and measures required. It will be seen that several of these items are not particularly specific to timber.

The aim of the inspection is to assess the overall integrity of a bridge and its capacity to support its statutory loading requirements. A bridge inspection should check:

- deterioration and decay in the timbers and surface treatments, e.g. water-repellent stains;
- deterioration in the metalwork, e.g. protective finishes on steel plates;
- lying water on deck/joints;
- wear and tear of components, mechanical damage;
- bulging at abutments/piers, loose fixings;
- rust;
- debris on the deck.

The site in the vicinity of a bridge should also be inspected. Unsafe trees or damage to the banks of a river, for example, could be potentially harmful to the structure.

Table 8.1 Common issues requiring remedial work

Problem	Remedy
Vandalism of smaller detachable parts, e.g. handrails	Replacement
Mechanical damage by cleaning equipment	Replacement, local repair
Graffiti	Cleaning, re-finishing
Surface finish deterioration	Re-finishing
Wearing surface deterioration	Replacement or repair
Loosening of connections and fasteners	Re-tightening
Loss od pre-stress in laminated decks	Re-tightening
Cracking and blistering of asphalt pavements	Re-paving

Figure 8.4 Dutton Horse Bridge, on the River Weaver Navigation in Cheshire. This is a unique Grade-II listed structure dating from the first decade of the 20th century. It is formed by mechanical laminating, using greenheart. In 1994–1995, skilled repair work was undertaken by British Waterways Technical Services, with advice from TRADA Technology.
A: the poor condition of some laminations that had previously been repaired with an inferior timber;
B: routing-out the decayed laminations using a guided milling cutter;
C: the renovated deck, using non-slip bitumen, with greenheart handrail posts;
D: the first of three coats of microporous water-repellent and fungicidal finish in the traditional white;
E: a decayed parapet zone before repair
Photos © Mr Raymond Harben, British Waterways

8.7 Inspection

HA Standard BD63 gives general guidance on inspection procedures for bridges. The particular aspects that should be considered in the inspection of timber bridges are included in the following sections.

8.7.1 Pre-inspection evaluation

A bridge inspection may be divided into three major steps: pre-inspection evaluation, field inspection and preparation of reports and records. A pre-inspection evaluation involves reviewing all the known information on the bridge, including original construction documentation, previous inspection reports and environmental information. These factors give an idea of the potential condition of the bridge, and may highlight special features or zones that should be closely examined during the field inspection.

8.7.2 Field inspection

Areas that may have been identified in the pre-inspection evaluation should be concentrated on first. The moisture content may be high, for example, and breaks in the finish or preservative layers are likely to provide an entry point for decay organisms. Regions close to connections and fasteners should be particularly carefully inspected (see *Figure 8.5*).

A walk across and around the structure and an examination of all of the easily accessible areas of the bridge will enable the inspector to observe the general features and look for obvious signs of deterioration or distress.

Figure 8.5 Inspecting a through-beam glulam pedestrian crossing bridge which lacks adequate protection – compare this with examples in Chapter 6 showing protective covers over beams – a deep-probe moisture meter reading of more than 30% indicates that soft rot decay risk is present
Photo © CJM

Changes in longitudinal or transverse deck orientation and levels could, for instance, indicate foundation movement. Drainage patterns should also be observed, together with the effectiveness of the deck and wearing surface in protecting underlying components.

Further methods that can be employed during inspection are summarised in *Tables 8.1* and *8.2*.

Based on the findings of such inspections, more detailed examination of both the substructure and the superstructure may be required.

8.7.3 Substructure inspections

These should start with a detailed visual examination of the abutments for signs of deterioration, mechanical damage and settlement. Trapped moisture and dirt is often clearly evident. Timber bank seats, wing-walls and sub-deck bracing should be examined for fractures. Bulging and seepage from earth pressure may be apparent. The tops of pile and posts should be inspected carefully.

Above the supports, the caps directly bearing the superstructure may have trapped debris and water run-off from the deck. These regions may require measurements and photographs so that improvements can be designed to prevent repetition. Connections into the cap, and horizontal fissures that have trapped water are critical. Signs of crushing should be investigated as these may indicate overloading or deterioration in the supported members. If piers and abutments are surrounded by water, even more specialised inspection is required.

8.7.4 Superstructure inspection

Often the first signs of trouble will be neglected surface finishes, or untreated surfaces that are in poor condition. While this situation is not necessarily disastrous, it does suggest a more thorough examination is required. Abnormal deflections, loose or missing decking and damaged connections are also symptoms of likely problems.

After the preliminary 'walk round', the underside of the superstructure is normally inspected first, since critical components may be obscured by the wearing surfaces and deck. Decay potential is highest where water may have passed through the deck, forming moisture at member interfaces, connections and fissures. Using a moisture meter, these regions should be tested thoroughly. Any signs of high moisture content or decay in these zones indicate the need for further investigation using one or more of the techniques summarised in *Table 8.2* and *8.3*.

Table 8.2 Simple methods and tools commonly used for preliminary inspections

Visual inspection	The simplest method of inspecting for overt deterioration. It needs to be performed by an experienced professional. Strong light is essential so that deterioration is clear. Visual signs include: fruiting bodies of fungi, obvious soft rot, localised surface depressions, staining or discolouration, insect flight holes, extensive deck wear and large deformations
Probing	Using a pointed probe, an inspector can locate deterioration near the surface, revealing excessive softness or lack of resistance. Care must be taken when interpreting the results since response varies according to the different species of timber
Moisture meters	The moisture meter is driven into the timber to varying depths, dependent on the type of probe. Generally larger hammer-probes are used on external structures such as bridges, giving indications at a greater depth within the section. By measuring the electrical resistance, the moisture content of the timber is determined at selected locations. Although this does not directly detect decay, the moisture inspection often leads to this diagnosis

Table 8.3 Summary of investigative techniques commonly used for detailed inspection

Drilling and coring	To detect internal decay, a special drill measures the resistance of a fine bit, providing an indication of voids and decreases in density throughout the section. Coring involves removal of a thin cylinder of wood from the section for visual examination of the signs of decay. It can also be used for a precise identification of the timber species, which may be important if the original documentation of the structure has been lost. This is a more damaging technique than detection drilling, but on a large structure, cores are relatively easily replaced with new material at a later stage in the repairs
Impact penetration tools	Instruments such as the 'Pilodyn' use a spring-loaded shaft to drive a hardened steel pin into the wood. The penetration depth for a standard impact gives a measure of the degree of wood softened by high moisture content and/or decay
Sonic evaluation	Using instruments designed and calibrated for the purpose, sonic waves are transmitted through the structure and detected by a receiver. The waves that are recorded by the detector are compared with the waves previously emitted, and any discrepancies can be analysed to give information on the internal structure of the wood

Particularly in unprotected bridges, decay may have occurred at the interfaces between the deck and the beams or arches. The regions below deck should thus be inspected for signs of deterioration and conditions likely to promote decay.

The upper surface of the deck should also be examined carefully. Subject to wear and abrasion from traffic, this may simply need replacing. The average value for the moisture content of a fully exposed timber deck will vary depending on the season, the timber species and the drainage details, but a figure beyond about 20% should give rise to concern, indicating that at the very least, cleaning and replacement of worn surfaces is required. Careful investigation should be carried out around fasteners, curbs and post attachment positions. If the deck has a non-slip surface, this will also need to be checked. Improvements in deck design or details should follow the protective design recommendations for new structures, discussed in Chapter 5.

8.8 Vibration testing

Uncomfortable vibrations should not occur, since they should have been foreseen in the original serviceability design (see Chapter 7). Otherwise, the resonance may make users feel insecure. Remedial measures are possible, but first the structure should be subjected to dynamic testing. Together with analysis and interpretation, this will indicate whether the natural frequency of the bridge is critical, in either a vertical or horizontal sense. In simple dynamic tests, typically, a relatively small mass is dropped from a controlled height and position to induce vibrational excitations, which are then measured by electronic instruments, linked to portable computers and analysers.

Similar techniques can be applied to estimate the damping characteristics of the structure, and to calculate the gross equivalent static stiffness of assemblies such as decks and beams. These types of investigation are normally carried out when fairly major changes to the structure are proposed. Sometimes they may be part of the process of conserving a historic structure that is to be updated for safe public access.

8.9 Reports and records

Accurate recording is essential to permit:

- identification of conditions that may limit the capacity of the structure or make it unsafe;

- a chronological record of the current structural condition, providing the basis for a further appraisal if/when conditions deteriorate;
- identification of current and future maintenance requirements.

Typical items to include in a proforma inspection report are:

- measured drawings, sketches of the structure – these may be based on original design drawings if available;
- a condition assessment of all parts of the structure, listed by components;
- a documentation of damage or decay – location and extent, indicated on diagrams and/or by labels and photographs;
- records of more unusual damage, e.g. softening of wooden decks or pavements by percolation of road salts (*Figure 8.6*);
- structural details that appear to have deviated from the record drawings and photographs;
- a narrative summary of the inspection findings;
- the recommendations for corrective action or routine maintenance.

8.10 Repairs

Maintenance and repair specifications can be drawn up only after a proper inspection. This requires the same level of thorough attention as a new-build design, including referring to the original code and associated standards (or in the case of historic structures, more modern, but sensitively interpreted standards). Prevention of moisture ingress is often the simplest method of reducing the risk of future decay, and this is likely to have the greatest pay-back in terms of improved life expectancy. Below about 20%, most fungal and insect growth will be prevented from recurring. Although modern timber bridges are beginning to be better protected (as discussed in Chapter 5), older structures are often poorer, and their protection by design may be inadequate. Hence consideration should be given to improvements.

Potential repair measures include:

- moisture ingress prevention and new sheltering/shielding (see Chapter 5);
- replacing members/components designed according to Chapter 5;
- in-situ preservative treatment for elements or complete components;

Figure 8.6 The famous Grade-I listed High Level Bridge, opened in 1849 and connecting Gateshead and Newcastle-upon-Tyne, England, was subjected to a major inspection, appraisal and maintenance programme over about ten years between 1999 and 2000.
A: all the timber beams and decking supporting the roadway and the separate footway were identified and closely inspected for strength grade equivalent and apparent damage or deterioration. Decay detection drilling was also undertaken. One of the effects identified was softening of the wood fibres through years of trickling precipitation enriched by industrial pollution and road salts.
B: the walkway supported by replacement timbers after the re-opening in 2009, viewing south towards Gateshead. Pedestrians and cyclists are permitted to use this restored route, with restricted use of the traffic lane by buses and taxis in a single direction. Client Mott MacDonald Ltd
Photo A: © TRADA Technology
Photo B: © Daniel Morgan

Figure 8.7 Hampden Bridge, Wagga Wagga, New South Wales, another important historic ironheart timber bridge designed by the highly respected civil engineer Percy Allen at the start of the 20th century. Rescued from proposed demolition, it is currently undergoing repairs. A and C emphasise the importance of the task of preparing methods statements for such major work; here, one of the truss spans is being winched off its supports on railway lines, for repairs at a nearby temporary workshop
Photos © Engineers Australia

- carpentry repairs – except for historic or vernacular designs, this will mainly be to the joinery;
- mechanical repairs, usually with fasteners, connectors and metalwork designed by codes for new-builds;
- adhesive repairs – for strengthening or stiffening.

8.10.1 Wearing surfaces

The addition of an asphalt wearing surface may provide a moisture barrier that not only protects supporting members, but also the deck itself. This protection may be increased when modern two-layer asphalt systems are placed on an additional membrane (see Chapter 6). For a laminated deck, re-waterproofing is often accompanied by re-stressing to restore the integrity of its plate action.

In-situ preservative treatments involve the application of preservative chemicals to arrest and prevent decay. Two types of treatment are commonly used: surface treatments to prevent infection of exposed timber and fumigants to treat internal decay.

8.10.2 Surface treatments

Treatments are either in liquid or paste form, using liquid tar oils or metal salt-based compounds. Pastes are available for treating vertical surfaces or openings. In the light of rapidly evolving regulations for safety of application, advice should always be sought from the treatment manufacturers. Boron-based compounds may be inserted into holes bored in previously damp zones of wood, as an additional precaution, but again it is inadvisable to depend exclusively upon this type of solution.

8.11 Mechanical repairs

Three general types of technique may be recognised:

1 member augmentation;
2 adding metal plates and reinforcements;
3 stress laminating.

All of these entail the use of steel fasteners, connectors and additional steel and timber components.

8.11.1 Member augmentation

This involves the addition of material to reinforce or strengthen existing members. The additional pieces, usually timber or steel plates attached with mechanical fasteners, increase the effective section and thus load capacity. Modern structural timber composites offer scope in such applications.

8.11.2 Metal plates and reinforcements

Fasteners and steel assemblies may be used to limit fissures or de-laminations in timber members. They are also useful to reconnect replacements. Components to which clamping and stitching methods may be applied include trusses and other structures with a large number of relatively small members and connections. Mechanically laminated bridges may benefit from similar techniques. Plates and other steel sections can also be used to attach replacements for decayed lengths and to effect moment-carrying splices.

Generally, plain carbon steel plates and standard structural steelwork sections are used. Suitable corrosion protection is essential, and this may generally follow guidance for new-builds. Work using appropriate grades of stainless steel is an alternative, and naturally this entails specifying matching fasteners, once again following the guidance for new projects.

8.11.3 Stress laminating

This technique was first introduced as an effective method for the mechanical repair of existing nailed laminated decks. The laminations are compressed with a series of high-strength steel rods applied transversely to the length of the laminations. These methods are described in Chapter 7. Re-stressing aims to compress the laminations laterally, and if carried out correctly, this greatly improves the load-distribution characteristics of the deck and reinstates a watertight under-surface, which should then be sealed with top layers of membrane and bitumen.

8.12 Adhesives for strengthening and stiffening

The formaldehyde family of adhesives that is widely used in timber engineering for new-builds (Chapter 4) require narrow gaps and very precise control of clamping pressure and curing conditions. This causes difficulty with in situ repairs, so selected formulations from the epoxy family are usually preferred. This is a wide-ranging group of civil engineering chemicals that have a track record of more than 60 years of successful use. However, they remain proprietary formulations, and an adhesive manufacturer's advice is essential. Correct identification of the exact aims of the project is essential in order to identify the appropriate materials specifications. This will guide those planning the work towards choice of suitable formulations with respect to aspects such as consistency (paste-like or thixotropic, for example) and also procedures with regard to surface preparation, cleanliness and moisture content.

Various epoxy formulations are available, some acting as bonding agents (principally adhesives) and others as grouts or fillers. The formulations differ significantly for the various functions, and specialist advice is essential. When correctly specified, mixed and applied, epoxy compounds form strong and durable bonds and surfaces. Vital measures include thorough preparation and cleaning of the surfaces of the timber that is to be repaired or stiffened. Also requiring close attention are the repair adherents, in the form of replacement timbers and connecting or stiffening plates, which can be metallic or non-metallic, e.g. glass fibre reinforced polymer (GFRP) (see the example below).

8.13 Bridge strengthening example: Tourand Creek Bridge, Winnipeg, Manitoba, Canada

8.13.1 Appraisal

Built in 1961 on Highway 59, south of Winnipeg, Manitoba, Tourand Creek Bridge is typical of many timber bridges that the highways services are required to strengthen. Although the 'stringer-decked' Canadian timber highway bridges of the era (from about 1930 to 1970) contained several of

Figure 8.8 Work was carried out under a protected and heated tenting arrangement. External ambient temperatures were around −20°C. The interior was heated to 15°C
Photo © D. Smedley Rotafix

the elements of modern timber engineering theory and practice, they were not built with stressed laminated decks, and did not anticipate the traffic weights and frequencies now experienced.

In 1999–2000, the authority committed C\$110 000 to a testing programme to evaluate this bridge, and to monitor the installation of GFRP pultruded, adhesively bonded reinforcements. The target, which was achieved, was to enhance the strength capacity of the bridge by at least 30%. If successful, some 575 additional bridges could be managed in a similar way.

The strengthening techniques entailed bonding GFRP rods into longitudinal grooves machined in the lower, or soffit, edges of the stringers. The selected epoxy adhesive was a slow-set, thixotropic, wide-gap-filling type, designated Rotafix CB10TSS, supplied by a specialist UK formulator and shipped to the site, and to the University of Manitoba for pre-testing, along with sets of 10 mm diameter GFRP pultruded rods. These are a special type, in which the glass fibres are embedded in an epoxy, rather than a phenolic matrix. This relatively low-stiffness matrix is highly compatible with the adhesive, thus reducing stress concentrations at the rod-to-wood interfaces.

Figure 8.9 Preliminary tests were conducted on half-sized (100 × 300 mm) timber sections. The edge-grooved slots, shown in these photos, were eventually rejected, as they were more difficult to cut in situ than soffit-edge grooves. Final tests were conducted on full-sized 200 × 600 mm stringers, with the grooves cut in the soffit edges, as shown in following illustrations
Photo © D. Smedley Rotafix

The testing, monitoring and installation work was a severe trial for the materials involved, as well as for the techniques applied. It entailed operating in extremely low temperature conditions, and bonding to timbers that were heavily impregnated with pressure-applied creosote preservatives.

8.13.2 Testing

Before work could start on the bridge itself, a significant testing programme had to be conducted at the McQuade Structures Laboratory of the University of Manitoba. *Figure 8.9* shows a half-scale reinforced beam, under flexural testing, while *Figure 8.10* shows a close-up detail of the type of wood found in the bridges, and the preliminary arrangement of GFRP reinforcing rods. The tests proved that GFRP rods would be a practical solution. Corrosion-resistance against road salts was another factor in their favour.

The full-sized bridge stringers were 200 × 600 mm section, a large size in solid Douglas fir, even by modern Canadian standards. The three-span, 23.3 m bridge, was chosen as a typical example of the highway structures that require strengthening. Consisting of two 6.4 m approach spans and a 10.5 m main span, it incorporates two standard stringer sizes used in many bridges of the type. Final tests were conducted on full-sized 200 × 600 mm stringers in the laboratory, before the actual installation work commenced.

8.13.3 Preparation

Deflection monitoring and level measurements were undertaken before the strengthening work began. *Figure 8.11* shows routed channels cut into the soffit edges of the stringers. Note the creosote that is still evident on the wood surfaces, within the grooves.

8.13.4 Execution

The work stages of the rod bonding process are shown in *Figures 8.12*, *8.13* and *8.14*. These show injection of the primer layer of thixotropic CB10TSS epoxy adhesive; insertion of the GFRP rods and their temporary attachment by staples.

Figures 8.12, *8.13* and *8.14* show the GFRP rods that were supplied from the United Kingdom and the remainder of the installation steps. When operations were finished, the bonded-in rods were encapsulated in two layers of cured epoxy resin. They were thus protected, both from the creosoted wood surfaces and from future potential corrosive agents, such as precipitation and road salt solutions.

Figure 8.10 As can be seen in this close-up, the timber was heavily impregnated with creosote. Assessing adhesive bonding to treated Douglas fir of this type was an important aim
Photo © D. Smedley Rotafix

Figure 8.13 The second layer of adhesive was injected around the rods
Photo © D. Smedley Rotafix

Figure 8.11 A primer layer and bed of CB10TSS adhesive was first injected into the clean-cut grooves
Photo © D. Smedley Rotafix

Figure 8.12 A hand-held stapling tool was used to retain the rods temporarily in position, whilst the first layer of adhesive cured
Photo © D. Smedley Rotafix

Figure 8.14 The second layer was smoothed to encapsulate the rods and leave a slight protective coating
Photo © D. Smedley Rotafix

8.13.5 Advantages and savings

These trials demonstrated installation advantages and, in addition, the sought-after strength and stiffness gains were attained. No heavy equipment was needed for the work, which took place in very cold weather, using hand-held tooling. There were no obstructions to the use of the bridge by the travelling public. Heavy vehicles were able to pass while the work was underway. The cost of performing the grooving, bonding and rod-installation operations included a sum of C$1500 per stringer for materials and work to completely bond-in the requisite set of rods. This led to the indicative overall saving per bridge of 15% over a replacement structure.

Figure 8.15 A large vehicle crossing the bridge while work was in progress
Photo © D. Smedley Rotafix

9 General case studies and final recommendations

9.1 Interpreting timber bridges

The world-wide revival of timber bridges has, not unexpectedly, led to differences in emphasis and approach in the various interpretations (*Table 9.1*).

This publication has been written:

- to inform all parties;
- to raise everyone's awareness;
- and to increase our expectations.

Table 9.1

Region	Emphases	Outstanding features
Nordic region	Clear differences between the countries in the region. Norway, with its maritime climate, chooses double preservative treatments as well as protective covers and ventilation. Sweden is closer to Germany in protective design; preservatives are avoided. Finnish techniques are intermediate – oil-based pressure treatment is accepted for decks, but elsewhere full use is made of covers, ventilation and surface treatments; practical advances have been made in the use of timber–concrete composite decks	Advances in large timber arches and trusses, both for lightly loaded structures and for main highway bridges – especially Norway and Finland. Inserted steel plate connections with plain dowels, manufactured by CAD-CAM
France	Utilisation of French-grown timber – preferably with a degree of natural durability but also protected by covers and ventilation. Emphasis on selecting material growing close to the sites – reducing 'wood miles'. Protection includes recommendations for surface finishes	Less of an inclination to use spruce than in Germany, even when fully covered. For the Pont de Crest, a pioneering supply chain environmental impact analysis was conducted which has further encouraged timber uses
Germany	Avoidance of chemical wood treatment, with high degrees of protective design, including metal covers and cladding over vulnerable surfaces and details. Acceptance of appropriate combinations of materials, e.g. timber + steel, timber + structural glass	Block laminating is well established by certain specialist suppliers, leading to very shallow side-profiles and advanced, curved-plan decks. Timber bridges are fully accepted by highway authorities
Switzerland	Technologically similar to Germany; highly specified designs with consequent likelihood of great endurance. Continuation of the covered bridge tradition, where appropriate – they are sometimes seen in Austria too	Advances in bonded-in rod technology began here, and were also investigated and used in Finland, then elsewhere. These are accepted in bridges in service class 2, i.e. protected situations
North America	The revival of timber road bridges was triggered by successful repair and upgrading projects, e.g. the introduction of stress laminating, first applied in repairs and then in new-builds. Today the United States has over 40 000 timber road bridges of short span, and these continue to be installed in Canada too; these programmes have had some influence in Mexico and Central America. Designs tend to be very utilitarian in appearance – the opposite end of the scale to Swiss bridges	Many covered bridges remain, although a significant number were lost before they became a recognised 'heritage'. Unfortunately, other types of great interest to industrial historians – e.g. uncovered Burr trusses – are neglected and being torn down
Southern hemisphere	Radiata pine from sustainable plantations is extensively available in Australia, Chile and New Zealand. Indigenous hardwoods are used in small quantities for special situations. Design of stressed laminated decks including use of radiata LVL well advanced by University of Technology Sydney and associated road authorities	Kerto LVL from Finland is somewhat harder to pressure treat, although not out of the question. New laminates probably less established at present in Antipodes. Historic trussed bridges – a range of unique types – now cherished and well conserved

In summary, the succinct advice is as follows:

- Select the timber species with great care, following the given protection strategy with its three strands.
- From the start, incorporate the chosen protection within the aesthetics of the bridge.
- Have very clear ideas on how to resolve the structural scheme, ensuring stability as well as strength and serviceability.
- Note the importance of the connections.
- Consult with potential timber supplies and manufacturers during concept design in order to avoid painful re-working.

In this way, designers should strive for *Firmitas*, *Venustas*, *Utilitas* – Vitruvius' three ideals, observed by Palladio in his four designs for timber bridges, published in *I Quattro Libri* in 1570!

In Chapter 2 we classified bridge structures as:

- beams
- cantilevers
- suspended structures
- arches
- trusses.

Explicit cantilevers are nowadays rare, although in *Figure 2.2* we have seen one contemporary example. Cantilevering is also still found within systems where continuous beams and triangulated frames are built – see, for instance, the Pont de Merle, Section 3.19. In the following, eight further examples are given, comprising brief descriptive case studies of an environmentally sensitive beam bridge upon which a carbon capture study has been made; an innovative suspended structure providing access for pedestrians and cyclists; three arched bridges (one for light vehicular traffic, one for cyclists and pedestrians and another in a rural location, for walkers only). There are also truss bridges showing two non-vehicular types and one very large trussed structure – a world-record span. These examples have been selected for the excellence of their design, and the reader should note how they illustrate the protective measures and structural integrity that we commend.

9.2 Pont de Crest: a contemporary environmental approach in central France

With all construction materials, their manufacture and transport inevitably entails the release of CO_2, and possibly other undesirable 'greenhouse gases'. However, through its source as wood from growing plants (trees), the unique feature of timber is that it has a positive carbon cycle leading to sequestration (*Figure 9.1*). Forests also play other significant roles in climate protection, including deterrence of flash flooding through soil stabilisation and air cleansing.

Through the process of photosynthesis, a tree's leaves capture this naturally occurring gas, and in the presence of rain water, use the energy from sunlight to produce essential plant tissues and protective substances. Hence, growing forests and woodlands lock up substantial volumes of CO_2. But this amazing cycle only occurs when the trees are vigorous and healthy, which is one of the fundamental arguments favouring 'economic scientific forestry'. This principle remains unchanged, but sometimes unfortunately overlooked, since it was first formulated more than one and a half centuries ago.

Timber structures may thus diminish, rather than worsen the overall environmental impact of a building project. Because structures such as bridges necessitate the use of certain amounts of 'CO_2-negative' materials such as steel and reinforced concrete, and also entail excavations and the movement of components, the net overall result of a project may not be totally 'carbon neutral'.

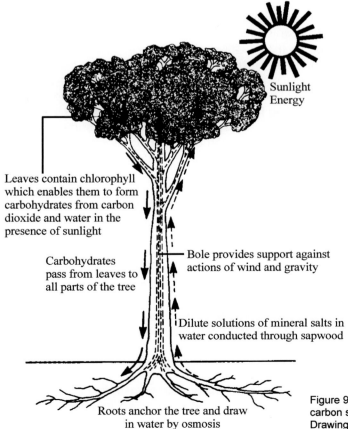

Sunlight Energy

Leaves contain chlorophyll which enables them to form carbohydrates from carbon dioxide and water in the presence of sunlight

Carbohydrates pass from leaves to all parts of the tree

Bole provides support against actions of wind and gravity

Dilute solutions of mineral salts in water conducted through sapwood

Roots anchor the tree and draw in water by osmosis

Figure 9.1 The growth functions of a broad-leaved tree, including carbon sequestration
Drawing © CJM – TRADA – *Hardwoods in Construction* TBL 62.

Figure 9.2 The Pont de Crest – a general view of the three-span structure crossing the River Drôme, which is sometimes far higher than in the season shown
Photos © CJM

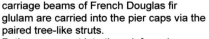

Figure 9.3 A: the loadings in the four main carriage beams of French Douglas fir glulam are carried into the pier caps via the paired tree-like struts.
B: these connect into the reinforced concrete caps via true pins formed as steel fabrications; the three materials are connected via invisible bonded-in threaded rods, inserted along the grain of the struts using an epoxy system adhesive
Photos © CJM

Nevertheless, it is likely to be significantly more benign than when structural timber is omitted, as shown by a detailed study of the Pont de Crest, undertaken by Bénédicte Sondaz, in association with the CTBA, and summarised below.

The overall approach: in this unusually thorough and detailed study, the overall approach was:

- analyse and quantify all the materials used in this light-traffic and pedestrian bridge;
- establish their individual ratios of CO_2 sequestered/emitted;
- examine the actual sources of the materials and their manufacturing locations, plus the distances to the bridge site;
- produce a complete CO_2 balance account for the finished bridge.

Description of the bridge: the Pont de Crest crosses the River Drôme at the town of Crest, which is to the south of Valance. The economy of the district is heavily dependent upon forestry and forest products industries, including sawmills and a large carton-manufacturing plant. Good-quality French-grown Douglas fir is available locally – it attains a high softwood strength class and has a fair degree of natural durability, reinforced by the protective design features of the bridge. Although not an indigenous species, this has been planted and grown for decades, resulting in a well-maintained and sustainable forestry cycle. The architectural inspiration for the bridge was from the trees themselves, reflecting in its form their branching structure and lightly curved limbs.

The main structure comprises four parallel glulam carriage beams, spaced 1.85 m apart, and crossing three main spans, giving a total length of 92 m – one of the longest timber road bridges in the country. These are supported on the tree-like struts that merge smoothly into fine-finish reinforced concrete pier cones. Two traffic lanes plus two footpaths provide the deck, which has in total an 8.5 m useable width. Sap-free Douglas fir laminations were used to manufacture the beams and struts, and other interesting technical features are included in the deck and the connections. Beneath an asphalt-sealed wearing surface (Chapter 6) is an oak-planked deck, followed by a sealing membrane and a structural diaphragm of cross laminated and bonded slices of sap-free Douglas fir.

The main beams are lightly curved with a parabolic rise. Four sets of paired 'W' struts span 4.2 m lengthways, and these are canted to conduct the vertical forces from the four beams into pairs of concrete pillars. These connections use the modern bonded-in timber rod method, with epoxy-system adhesives. The carriage beams gain lateral stability from sets of lateral 'K' struts at 4.0 m longitudinal spacing.

The influences of procurement were also assessed, and the stages of production, locations of plant, and road travel factors were calculated.

9.2.1 Conclusions
- For each of the five principal structural materials in the Pont de Crest, the individual ratios of CO_2 sequestered/emitted have been investigated and calculated using realistic data based on the actual project.

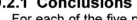

- Normalising the balance figures, including all aspects of their final presence and volumes in the structure:

  > steelwork + 0.67; reinforced concrete piers + 0.47;
  > foundations + 0.27; timberwork − 0.41; net + 1.00.

  where a positive value shows CO_2 released into the atmosphere and a negative figure the CO_2 locked in, or sequestered.
- Quantitatively, the net CO_2 balance for the bridge is 154 tonnes; without the ameliorating effect of the timberwork it would have been at least 300 tonnes, because not only did it deduct from the sum, but a steel or reinforced concrete superstructure and deck would have been necessary.
- For all five of the main structural materials, transport affects the carbon balance in an undesirable manner, but this is not a major contributor to the total − nevertheless, the use of local timber supplies and manufacture should be encouraged and would add further to the benefits.
- To attain a neutral carbon balance, for every cubic metre of reinforced concrete used in a similar bridge, $1.25\,m^3$ of timberwork needs to be included − this should be achievable in future projects, particularly for lightly loaded pedestrian and cycle bridges.

9.3 Punt la Resgia: a modern covered arch design in Switzerland

Around 1990 several designs of a similar type were built across the river Inn, in the Resgia Canton of Switzerland, typifying the advances made at that time by combining modern timber engineering materials and techniques with essentially a very traditional national form of bridge. Their historical precedents include the pure arches used by Hans Ulrich Grubenmann over the river Limmat at Wettingen at the end of the 18th century (see Chapter 2).

Those earlier covered arches used thick, bent and mechanically restrained laminations, whereas nowadays curved glulam is employed. Appropriate timber species are chosen for different parts of the timberwork. For instance, oak is used where high axial strength is required, larch is designated for cladding – subjected to significant weathering – and the less durable spruce is only employed in the fully protected parts. For the Resgia bridges, a portal system provides lateral stability, avoiding a 'cluttered' architectural appearance, while the attractive low-pitched roof is given hipped-gabled ends and a slight lengthwise flare on plan.

This bridge over the river Inn at Ramosch links the community to another location called Waldungen, where for centuries the forests have been carefully managed and are selectively harvested. The bridge not only provides a general thoroughfare, but also facilitates the transport of logs, supporting the local economy. The Swiss National Standard SIA 160 loading was applied – for secondary access roads, this implied a gross weight of 36 tonnes, with two six-tonne axles. At the sides of the deck, a 1.0 m wide zone is raised as a pedestrian walkway, while the useable deck width is 4.0 m. Beneath the portals, the clearance is 4.60 m – sufficient for loaded logging trucks.

Figure 9.4 Punt la Resgia – a modern covered road bridge with glulam arches
Photo © JPM TRADA

Structural system: With a rise of 6.8 m and a span of 43.0 m, the pair of two-pinned glulam arches form the basis of the design. Their section is uniformly 220 × 1730 mm, while to facilitate prefabrication and site erection, their crowns are rigidly spliced using inserted steel plates and dowels. Such arches clearly require lateral stabilisation, and this was cleverly attained by a method that also provided neat transom hangers and roof portal posts. These high-duty verticals straddle the arches, and are formed from a combination of oak members of 160 mm square section with a structural timber composite of 200 mm square. The transoms are 5.62 m long, and their end grain is totally protected by means of local downward extensions of the parapet cladding – a good example of the thoughtful protective detailing that is a modern-day essential. The wearing surface of the bridge is a frost-proof mesh-reinforced concrete system. This is taken over the transoms by means of a 'permanent form-work' of timber, with an intermediate membrane. At the tops of the vertical portal posts and transom hangers there are pitched-cranked lateral beams, together with longitudinal eaves beams. The latter connect with a symmetrical horizontal cross-bracing for the wind forces that are taken down to the foundations at the end-portals. At the bridge entries, the timber stanchions are raised above reinforced concrete pillars, ensuring good impact resistance and ample drainage, again away from the vulnerable end grains. The pitched and gabled roof is clad with a copper–titanium–zinc alloy, supported on timber boarding, and its edges are neatly trimmed and protected with a curved wooden down-stand.

Figure 9.5 Punt la Resgia – the view through the entry, showing the gabled and flared roof, the roof bracing, plinths, composite hangers, arches, footpath and roadway
Photo © JPM TRADA Technology

9.4 Toijala, Tampere: a bridge with distinctive trusses for cyclists and pedestrians – Finland

This is a cyclist and pedestrian crossing bridge along a section of the Helsinki-to-Tampere motorway. With funding from the National Development Agency and Wood Focus Oy, it was developed in association with the nearby Wood Information Centre at the Tampere University of Technology. In addition to providing a safe crossing for students and other members of the public, the objective was to generate a series of contemporary references showing good implementations of contemporary timber engineering.

Figure 9.6 The cyclist and pedestrian-crossing bridge at Toijala, near Tampere, Finland
Photos © Puu/Finnish Timber Council

160

In this way, an unusual form of trussed bridge was developed. The central segment of the top chords act in compression, while the outer chord members are in tension. Tensile steel rods are also included, creating a reasonably light overall appearance.

Loadings are a uniform $4\,kN/m^2$, plus provision for service vehicles with axle loads of 80 kN and 40 kN over a wheel-base of 2 m. Each of the two spans measures 30 m, and the continuous deck has a width of 4.5 m.

The top chord comprises twin glulam members of $180 \times 600\,mm$ cross-section, and the same section is used for the straight bottom chords. Triple sections of $115 \times 400\,mm$ make up the spaced glulam struts that support the upper elements. The deck consists of 140 mm thick glulam pine slabs, which have been pressure preservative treated after shaping, using creosote. The remaining glulam components were treated under pressure with water-borne salts and then dried and lightly machined prior to laminating.

Figure 9.7 A striking and unusual form of trussed bridge
Photo © Puu/Finnish Timber Council

Figure 9.8 The deck, formed with 140 mm thick pressure-treated creosoted pine glulam slabs
Photo © Puu/Finnish Timber Council

Figure 9.9 The general view of Ollas Overpass, from the embankment – timber lamp standards are included
Photo © Puu/Finnish Timber Council

9.5 Ollas, Finland: a pedestrian and cyclist flyover using classic glulam arches

This pedestrian and cycle bridge was constructed in 1997 and is situated in Espoo, about 20 km west of Helsinki. It has an alignment below the natural ground level, and advantage is taken of the hard rock directly beneath the foundations. This enables horizontal thrust to be transmitted straight into the ground. The static system presumes two-hinged arches, which are located each side of the deck. Thin neoprene bearings support the timberwork.

The span of each arch is approximately 33 m, while the horizontal clearance of the deck is 4.0 m.

Vertical clearance between the cross beams and the road surface is about 5.2 m. This is half a metre more than the minimum requirement, because the design allows for a transverse crash force of 300 kN onto the support and bracing trestles.

The plate deck comprises pairs of glulam slabs, which are interconnected laterally. These and other paired elements are connected using concealed toothed plate connectors. The glulam parapet includes steel netting below the handrail.

In considering the aesthetics, the designers paid considerable attention to details such as the pairing of the arch and cross beam elements, with concealment of bolt ends by wooden plugging, and sloped blocking-in above the arch gaps.

The protection strategy is a combination of detailing to shed water and attain free air circulation, plus two forms of wood preservative treatment. The deck slab and the transverse trussed trestles are of creosote-impregnated glulam, since these are unlikely to be inadvertently touched by the public, while the arches, railing posts and handrails use salt-impregnated timber, which has been finished with yellow and blue microporous paint.

Figure 9.10 A: the deck support trestles, designed for an impact force;
B: showing some of the attractive details, these also have an important practical function in improving protection
Photos © Puu/Finnish Timber Council

9.6 Flisa Bridge: a record-breaking trussed road bridge in Norway

With a main span of 71 m and an overall length of 180 m, the Flisa Bridge crosses the Glomma River in Hedmark County, Norway. Since its completion in 2003, this has been the largest timber road bridge in the world. Its width is 10 m, and its truss elements are up to 600 × 710 mm in section, broken down into 28 m lengths for prefabrication and delivery.

Like the other large Norwegian road bridges such as those at Evenstad and Tynset, the Flisa Bridge uses multiple slotted-in steel plates and dowels for the connections, as discussed in Chapter 8. Double preservative treatments were applied, injecting the individual laminations with copper salts, then drying them, following the normal glulam manufacturing processes, and finally applying a high-loading creosote treatment to the cut and slotted elements. Because of these measures, together with the 'durability-by-design' approach (see Chapter 3), the bridge is expected to have a life of more than 100 years – a longevity in excess of the steel bridge that it replaced!

The old bridge comprised three spans of rather conventional steel trusses, built in 1912, with only one lane. Due to its deterioration in strength, an eight tonne weight restriction had been applied. In Norway, a major factor in the weakening of such steel bridges has been found to be corrosion through road salting, whereas with timber structures, salt actually helps with the preservation.

The new bridge has two lanes, with a 13 tonne capacity for vehicular traffic, plus a pedestrian walkway. Inspections of the existing stone masonry piers and timber-piled foundations showed both to be in good condition. With a serious flood record in the Glomma valley, the high-water mark, plus the icy conditions in winter, became critical parameters in the design. Hence the deck was raised slightly through the introduction of stronger pier caps, while the lower existing stonework was retained. New abutments were built, and at the piers, further steel sheet piling was added. These modest changes contributed significantly towards the overall project economy.

Another consideration relating to the site profile was the location across the river of the existing piers and abutments. The centre span and the

Figure 9.11 The general view of Flisa Bridge, with construction nearing completion; note the steel portal bracing system at the entry to the foreground truss, just above the intermediate pier
Photo © CJM

crossing from one of the banks measure 56 m, but the gap to one of the piers from the other bank is over 70 m. Hence an asymmetrical structure was required. Taking account also of the cross-sectional performance needed for the bridge, the design team adopted a Warren truss through-bridge, in general appearance not unlike the old structure, but having the wider and heavier load capacity, and of course made of glulam.

Through the low mass of timber, and the fact that its fabrication costs were cheaper than steel, the new design saved about US$1 000 000 compared with the alternative of a replacement in steel. Maintenance costs were also evaluated, and it was decided that a natural, dark, creosoted finish on the main structural glulam members would be acceptable. This avoided the costs and the environmental hazards of having to re-paint a steel bridge at regular intervals.

The large central glulam trusses are cantilevered out by 17 m beyond the piers, so as to reduce the longitudinal forces on the piers, and to create convenient side spans. The main trusses have bowstring top chords, while the larger side span has a pair of concave trusses, giving an overall aesthetic that meets with local approval, in recalling the original bridge. Some elements of the new bridge, in fact, are of steel working in partnership with timber. This was because in some positions, greater slenderness was required than possible with timber. An example is in the cantilevered U-frames that take the deck out to a plan position well beyond the side faces of the piers. Such choices often make these large, mainly timber road bridges acceptable in appearance, and solve special structural difficulties. The unity in safety format given by the Eurocodes suite makes bridge designers conformable with this combined-materials approach.

Fitted above all of the main timber elements is a protective cladding of copper alloy. Air-gaps are included beneath this cladding, and any water that does trickle through reacts with the copper to form salts that are effectively a wood preservative. Since 1996 moisture monitoring had been carried out on the Evenstad Bridge, and in a six-year period, the average moisture content of the timbers of this bridge (a similar type of double-treated glulam) had not exceeded 12%. In accordance with the recommendations of BS EN 386, the glulam thickness is restricted to 30 mm for external service conditions. As well as ensuring maximum uniformity of bond-line pressure, this also assists with the penetration of the salt preservatives in the first stage of the treatment process.

The Flisa Bridge includes what has now become a conventional stress laminated pine deck, pressure-treated, with an asphalt wearing surface. See Chapter 6 for more discussions of such deck structures, whose design principles are given by BS EN 1995-2.

The design calculations were made in accordance with Eurocode 5 and other compatible codes and standards. The Nordic Timber Bridge Programme included specific research projects such as fatigue tests on timber bridge connections, and investigations into various forms of deck. Together with the practical experience gained with other large road bridges such as those at Evenstad and Tynset, the confidence was gained to make the record-breaking Flisa Bridge possible. At 2003 prices, its total cost was less than £15 000 000. Only seven months after the steel bridge that it replaced was closed, the bridge was opened on time, with erection only having taken three months.

9.7 Almere pylon bridge: a pedestrian and cycle crossing in the Netherlands, using innovative timber engineering

A new form of timber that can achieve excellent results both structurally and architecturally has been demonstrated at a site in the vicinity of Amsterdam (*Figure 9.12*). The structure has load-carrying elements of 'block laminated timber', or 'Brettschichtholz' as it is known in German. This is twice-laminated material, manufactured on the same principles as glulam, but producing a more sophisticated outcome. With its technically complex and attractive high-level supports visible from considerable distances around the site, the pedestrian and cycle bridge crosses a heavily trafficked road from the city to an area that is protected for its natural habitat.

The city planners were seeking an architecturally outstanding solution, and the idea of a cable-stayed 'pylon-bridge' emerged. The deck was to be sinuous in plan, so the choice of the material was significant. The concept design seemed to be uneconomical in certain materials. For example, in reinforced concrete, the high self-weight of such a structure was infeasible, while with steelwork, the curved geometry was very expensive. Thus the planners quickly accepted a timber design using 'block laminated' material.

The sustainability credentials of the new generation of timber structures was an added attraction to the use of this material – an aspect greatly appreciated in the Netherlands. Smaller bridges in the region are often constructed using tropical timbers, but this new project extends the range of materials options available to bridge owners, exploiting the technical prowess of the fabricators. Similar bridges have been successfully established in Germany, but elsewhere, such solutions are quite rare, so the designers and timber producers were delighted to have the opportunity to disseminate their experience.

Behind the imaginative bridge form lies careful static concepts – bracing the suspended deck from twin offset and canted towers expresses the

Figure 9.12 In Almere, close to the coast of the Netherlands, this new cable-stayed timber 'pylon bridge' was completed during 2008. It demonstrates a new technique known as 'Brettschichtholz', or block laminated timber
Photo © F. Miebach/Schaffitzel

force relationships with transparency and clarity. Hung from the two round-hollow-section stayed steel masts is the block laminated timberwork. The length of the bridge is 75 m, with a 3 m deck width. The pylon heights are 18 m above the foundation level, while the offsets of the 'S-shape' plan are each about 9.5 m.

Using the stabilised pylon systems together with their bracing stays and steel transoms, under the main load cases, only normal forces are carried in the principal timber structures. Each curved glulam deck member has a thickness of 560 mm and a width of 2200 mm, with a free span of approximately 48 m.

These elements were block-bonded in sets of three, using special jigs developed by the producers. With a radius of approximately 100 m in plan, the deck elements are also pre-cambered upwards by about 900 mm in each span. The actual wearing surface of the deck is formed with reinforced concrete plates. These units are sealed with a high-grade membrane over the timberwork. Additionally, sheet-metal gutters are provided, formed with high-grade stainless steel.

The bridge's structural design complies with the new DIN 1074 'Timber Bridges', which places strong emphasis on protective features to avoid attack though moisture effects. Hence the design life, designated according to the regulations as 80 years for a bridge of this type, is targeted just as it would be for a bridge of concrete or steel.

The construction gained considerable advantages through prefabrication, since all of the major components were built in the dry in workshops, starting in January 2008, when outdoor conditions were well below freezing. Rapid erection was also achieved because of the high degree of pre-assembly, despite continuing adverse weather conditions.

Figure 9.13 A sinuous deck plan was an important aspect of the design
Photo © F. Miebach/Schaffitzel

 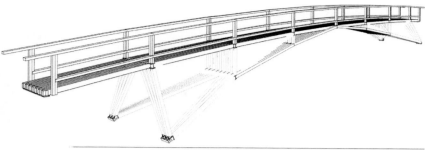

Figure 9.14 An earlier footbridge was swept away by floods, so a greater clearance is provided beneath the arches of this masterful new piece of timberwork
Photo and drawing © J. Anglade

9.8 Passerelle d'Ajoux

An elegant arch in locally sourced sweet chestnut, providing access for ramblers in the Ardèche.

Within the Ardèche, the Vallée de l'Auzène has numerous footpaths with a high reputation for natural beauty, and not only is an important route restored by this sensitive design, but the structure itself enhances the experience. The light, slender design is unobtrusive; its graceful arches are in tune with the large rounded boulders of the stream bed; and the natural landscape with sweet chestnut woods is all around.

The use of chestnut has special technical implications. When free from sapwood it is fully durable – however, it is really only possible to find in relatively short lengths and small sections. But in this case, this was by no means a restriction for the designers. Demonstrating the use of coppiced sweet chestnut is an important lesson for the United Kingdom too. There are abundant quantities in several UK regions, including the Welsh borders and the Weald of Kent and East Sussex, where there is a sawmill and end-jointing plant specialising in its use. But currently it is used insufficiently to support the necessary woodland conservation.

In the Ardèche, the local context was fully recognised by the mayor's office of Ajoux and Issamoulene, where the district councillors were especially keen to employ their abundant local resource. The engineer Jacques Anglade, who has an affinity with the region and a particular respect for its forest products, confirmed the moisture resistance provided by the high tannin content of this timber, and instinctively responded with a very appropriate design.

Figure 9.15 Designed for manual assembly on-site after prefabrication of the timber engineered components
Photo © J. Anglade

Figure 9.16 A lightweight structure using short lengths and small sections of sweet chestnut
Photo © J. Anglade

The steep rocky terrain meant that erection techniques needed to be light and flexible. There was no question of moving heavy lifting plant into the site. Only hand-lifting, aided by Tirfor winches, cables and pulleys would be suitable. The opportunity to use short, straight lengths of a light-sectioned timber was seized. This also suited the woodworking capabilities of local carpenters with small workshops.

A cross laminated deck and a moment-fixed parapet crown the eight individual segmental arched frames, all in chestnut, but the frames are laterally stiffened by small discrete sections of electro-galvanised steel that cross the nodes. Triangulation takes a series of seven point loads, collected at the transom positions, down onto the arch frames. The latter are joined into a simple shell by means of lateral–vertical braced diaphragms, while the arches are given additional overall lateral stability from the flare that is built into their lowest segments, where the thrust passes into the stonework.

Lightweight and neatly fitting corrosion-resistant metal node connections also characterise the skilled timber engineering introduced by Anglade. These operate in a manner closely matching the idealisation that was provided by the initial three-dimensional computer modelling of the scheme. Not only this, but these connections were perfect for the manual assembly of the light frames that were carried into the woods in small sub-assemblies and individual prefabrications.

Figure 9.17 The structure seen below deck
Photo © J. Anglade

Figure 9.18 The Journal *Sequences Bois* notes that the small straight lengths of slender section echo the rapid torrents of water, but to an English eye, this small masterpiece also seems a classic of French Baroque design, and Jean-Rodolphe Perronet would certainly have approved!
Photo © J. Anglade

9.9 Tharandt Forest Botanical Garden: tree-top walkway

A municipality in Saxony, Germany, Tharandt is situated on the Weißeritz, nine miles southwest of Dresden, on the Dresden–Reichenbach railway (*Figure 9.19*). It has the oldest academy of forestry in Germany, founded in 1811 by Heinrich Cotta (1763–1844). There are strong historic connections both with the origins of tropical forest conservation, and with early collaboration between German and Oxford University foresters in this field.

Tharandt is a favourite summer resort of the people of Dresden, one of its principal charms being the magnificent beech woods in their surroundings. Tourists are also attracted to the area, and the forest garden is an important focal point. As part of the reconstruction project, a tree-top walkway has been built to connect two zones of the gardens that were separated by a busy road, the Freiberger Straße S194. The grace and distinction of the walkway amongst the treetops actually invites spontaneous visitors to the gardens. It also serves a very practical function for the gardeners and maintenance staff. For this reason, it was designed to permit small tractor and trailer loads as well as pedestrians, crossing unimpeded above the main road.

Over recent years, tree-top walkways have become recognised as an effective way of bringing people of all ages back into contact with woodlands and forests. Centuries old as a practical measure, they were first reintroduced in modern times for the serious study of tropical forests, but temperate woodlands too are vital for our health and well-being. Visitors can simply take a leisurely walk amongst the crowns of the trees, or they can bird-watch, study plant–insect relationships or just look more closely at the formation, colours and shape of the different leaves and branches of various species. Rather than being constructed completely from steel, this particular tree-top walkway actually makes direct use of the renewable resources of timber.

The design began in 2001 with an architectural students' competition at the Technical University, Dresden. In actually being built, there was excellent collaboration between the professors in the faculties and departments of architecture, civil engineering and surveying. The

Figure 9.19 Views of Tharandt, in Saxony.
A: looking from the castle towards the Schloitzbach Valley;
B: the Forestry Academy
Photos © Norbert Kaiser Wikimedia Commons

ultimate result was similar to the original prize-winning design of the two students, André Dreßler and Kathrin Gädeke, using bi-curved forms on round-section tripod stilts. This was facilitated by the innovative technology of prefabricated block laminating components brought about by the timber engineers Schmees and Lühn, of Fresenburg, Lower Saxony.

Some of the main constructional data are shown in *Figure 9.20*, which shows views of the multi-curved deck wending through glades and forest canopy. The block laminating process achieved relatively tight curves – in some cases down to a 16.25 m radius. After completion, as a training project for surveying students, measurements were made to assess whether residual stresses or long-term changes in curvature were induced in the assemblies, and these tests had reassuring results.

A rapid erection process during the autumn/winter of 2003–2004 proved the precision of the computer-assisted prefabricated construction, with an on-time opening ceremony taking place at Easter 2004. Since then, there have been good experiences in-service, the only subsequent maintenance measure having been a decision to apply a surface finish to the timber handrails.

Figure 9.20 The walk-way winds through the glades and treetops for 117 m, crossing 15 spans of 7 m and one of 12 m across the main road. Its deck is 2500 mm wide, and it is made from factory-prefabricated block laminated sections, comprising 220 mm thick vertically laminated main elements, surmounted by 40 mm thick Kerto LVL, a layer of structural plywood, a waterproofing membrane and a two-layer polymerised asphalt
Photo © Schmees & Lühn

Figure 9.21 A deck-level view, showing the steel and timber parapet – this format is readily mounted on the sides of a block laminated deck with adequate moment stiffness. Fine stainless steel wire mesh netting provides adequate security in a region where the deliberate dropping of objects onto the carriageway below is an unlikely occurrence
Photo © Schmees & Lühn

Notes

1 Benefits of timber bridges

1 Aasheim, E. *Development of timber bridges in the Nordic countries.* Pacific Timber Engineering Conference, Rotorua, New Zealand, 1999.

2 Kleppe, O. and Dyken, T. *The Nordic Timber Bridge project and the Norwegian approach to modern timber bridge design.* Innovative Wooden Structures and Bridges, Lahti, Finland, IABSE, 2001.

3 Berthellemy, J. (ed.) *Les ponts en bois : comment assurer leur durabilité.* Sétra, 2006.

4 Zeitter, H., Schwaner, K. and Damm, H. *Developments in wooden bridges in Central Europe.* Innovative Wooden Structures and Bridges, Lahti, Finland, IABSE, 2001.

5 Ritter, M. A. *Timber bridges: design, construction, inspection and maintenance.* US Department of Agriculture, Forest Service, EM 7700-8, 1990.

6 Duwadi, S. R. and Ritter, M. A. *Timber bridges for the 21st century: a summary of new developments.* Innovative Wooden Structures and Bridges, Lahti, Finland, IABSE, 2001.

7 Taylor, R. J. and Keenan, F. J. *Wood highway bridges.* Canadian Wood Council, 1992.

8 Crews, K. *Development of high performance timber bridges in Australia.* Innovative Wooden Structures and Bridges, Lahti, Finland, IABSE, 2001.

9 Bier, H. *Our timber engineering heritage and a sustainable vision for the future of timber structures.* Institution of Professional Engineers New Zealand, Vision 20/20 Insight, Christchurch, 2008.

10 Anon. *EFORWOOD – TOSIA: a tool for sustainability impact assessment of the forest-wood chain.* Skogforsk, 2009.

11 www.dietrich-ingenieur-architektur.de/IA-1-start-GER.htm (accessed 28 October 2010).

2 The evolutionary development of the timber bridge

1 Watson, Wilbur J. *Bridge architecture.* William Helburn Inc., 1927.

2 Peters, T. F. *Transitions in engineering: Guillaume Henri Dufour and the early 19th century cable suspension bridges.* Birkhäuser, 1987.

3 Vittore Carpaccio. *The patriarch of Grado heals a man possessed by a devil.* Gallerie dell' Accademia, c.1494.

4 Giovanni Antonio Canal (Canaletto). *View of the entrance to the Arsenal.* Private collection, 1732.

5 Pryor, F. M. M. *Flag fen: life and death of a prehistoric landscape.* Tempus Publishing, 2005.

6 Maggi, A. and Navone, N. (eds) *John Soane and the wooden bridges of Switzerland: architecture and the culture of technology from Palladio to the Grubenmanns.* Archivo del Moderno Accademia di architettura, Mendrisio Università della Svizzera italiana and Sir John Soane's Museum, London, 2003.

7 O'Connor, C. *Roman bridges.* Cambridge University Press, 1993.

8 Milne, G. *Timber building techniques in London, c.900–1400.* London & Midland Archaeological Society, Special Paper 15, 1992.

9 Meyer-Usteri, K. *Timber bridges in the upper Aargau in Canton Berne.* Kantons Bern, 2004.

10 Nelson, L. H. *The colossus of 1812: an American engineering superlative.* American Society of Civil Engineers, 1990.

11 www.tfguild.org/news/kickingbridge2004.html (accessed 25 September 2010).

12 Binding, J. *Brunel's Cornish viaducts.* PAtlantic Transport Publications, 1993.

13 Booth, L. G., and Booth, V. Timber railway bridges in England in the period 1835–1860: their structural forms and contemporary lithographic illustrations. *Journal of the Institute of Wood Science*, 14, 1, 1996.

14 www.betterpublicbuilding.org.uk/finalists/2004/moyviaduct (accessed 26 September 2010). *Timber beam bridges: study of relative heritage significance of RTA controlled timber beam road bridges in NSW*, Roads and Traffic Authority, New South Wales, 2000.

15 Usuki, S. and Komatsu, K. Two timber road bridges. *Structural Engineering International*, 8, 1, 1998.

16 Ochsendorf, J. *Engineering in the Andes mountains: history and design of Inca suspension bridges.* Library of Congress webcast, 8 December 2005 (accessed 24 September 2010).

17 http://atlasobscura.com/place/root-bridges-cherrapungee (accessed 24 September 2010).

18 DeLony, E. *Context for World Heritage bridges.* Historic American Engineering Record. National Park Service, ICOMOS and TICCIH, 1996.

19 Noltie, H. J. Snapshots of China: George Forrest's expeditions. *The Garden,* 121, 5, 1996.

20 Grob, A., Krähenbühl, A. and Wagner, A. *Survey, design and construction of trail suspension bridges for remote areas.* Swiss Center for Appropriate Technology, 1983.

21 www.tech21.ch/archiv – item 37/2005 (accessed 26 September 2010).

22 Conzett, J. The Traversina Footbridge. *Structural Engineering International,* 7, 2, 1997.

23 Wells, M. *30 Bridges.* Laurence King, 2002.

24 Paolillo, G. G. M. *Il ponte di Trajano sul Danubio presso Drobeta nella Dacia Inferior* [Trajan's Bridge on the Danube at Drobeta in the Dacia Inferior]. Bollettino Ingegneri 12, Collegio Ingegneri, Firenze, 1999.

25 Popa, N., Bancila, R. and Florea, S. *Trajan's bridge over the Danube at Drobeta Turnu Severin.* International Bridges on the Danube, Vienna–Bratislava–Budapest. IABSE, 1992.

26 Sitwell, S. *Bridge of the Brocade Sash.* Weidenfeld and Nicholson, London, 1959.

27 Shen, W. and Liu, J. *Chinese rainbow bridges.* World Conference on Timber Engineering, Vol. I, Lahti, 2004.

28 Ruddock. T. *Arch bridges and their builders 1735–1835.* Cambridge University Press, 1979.

29 Booth, L. G. The development of laminated timber arch structures in Bavaria, France and England in the early nineteenth century. *Journal of the Institute of Wood Science,* 5, 5, 1971.

30 Müller, C. *Holzleimbau: laminated timber construction.* Birkhauser, 2000.

31 Chugg, W. A. *Report on a visit to Switzerland to inspect glued laminated timber structures over ten years old.* TRADA Technology, 1962.

32 Wilson, T. R. C. *The glued laminated wooden arch.* US Department of Agriculture, Technical Bulletin 691, 1939.

33 Bell, K. and Karlsrud, E. *Large glulam arch footbridges: a feasibility study.* Innovative Wooden Structures and Bridges, Lahti, Finland, 2001.

34 Heyman, J. Palladio's wooden bridges. *Architecture Research Quarterly,* 4, 1, 2000.

35 National Society for the Preservation of Covered Bridges. www.coveredbridgesociety.org (accessed 9 July 2009).

36 Pierce, P. C. *et al. Covered bridge manual.* US Department of Transportation, Federal Highway Administration, Report No. FHWA-HRT-04-098, 2004.

37 Seymour, G. D. Ithiel Town. *Dictionary of American Biography Base Set,* American Council of Learned Societies, 1928–1936.

38 Luggin, W. E. and Luggin-Erol, K. Historic timber bridge system, Böheimkirchen, Austria. *Structural Engineering International,* 3, 2002.

39 www.rta.nsw.gov.au/cgi-bin/index.cgi?action=heritage.show&id=4300155 (accessed 26 September 2010).

40 O'Connor, C. *Spanning two centuries.* University of Queensland Press, 1985.

41 Ekeberg, P. K. and Søyland, K. New Flisa Bridge in Hedmark County, Norway. *Bridge Engineering Journal,* 158, BE1, 2005.

42 www.vebjørn-sand.com/thebridge.htm (accessed 28 September 2010).

3 Durability and protection by design

1 Dyken, T. and Kleppe, O. *The Norwegian approach to modern timber bridge design.* www.balticroads.org/conference25/files/kleppe-o2.pdf (accessed 28 September 2010).

2 www.balticroads.org/conference25/files/kleppe-o2.pdf (accessed 28 September 2010).

3 Schwaner, K. *Protection and durability of wooden bridges.* Innovative Wooden Structures and Bridges, Lahti, Finland, 2001.

4 Sengler, D. *Dokumentation und Ermittlung realitätsbezogener und bauartspezfischer Unterhaltskosten von Holzbrücken.* Forschungsbericht Deutsche Gesellschaft für Holzforschung, 1986.

5 *Timber: fungi and insect pests.* TRADA Technology, 2004.

6 www.wood-protection.org/restricted/Preserving_confidence_in_timber.pdf (accessed 28 September 2010).

7 Evans, P. Emerging technologies in wood preservation. *Forest Products Journal*, 53, 1, 2003.

8 Hislop, P. *External timber cladding.* TRADA Technology, 2007.

9 Davies, I. and Watt, G. *Making the grade: a guide to appearance grading UK grown hardwood timber.* Arcamedia, 2005.

10 *Pont de Merle, symbiose avec le site.* Séquences Bois, No. 35, CNDB, April 2001.

4 Materials

1 *Glued laminated timber: an introduction.* TRADA Technology, 2003.

2 Mettem, C. J., Gordon, J. A. and Bedding, B. *Structural timber composites design guide.* TRADA Technology, DG1, 1996.

3 *Structural glued laminated timber: design essentials.* Glued Laminated Timber Association, 2010.

4 *Structural use of hardwoods.* TRADA , 2003.

5 Mettem, C. J. and Richens, A. D. *Hardwoods in construction.* TRADA Technology, 1991.

6 *Timber strength grading and strength classes.* TRADA , 2006.

7 Booth L. G. *Two early uses of glued laminated timber in buildings in Britain: Henry Fuller's Rusholme Road Congregational Sunday School, Manchester (1864–1963) and Lower Clapton Congregational Church, London (1863–1931).* Department of Civil Engineering, Imperial College, Report TE 2/93, 1993.

8 Kreuzinger, H. *Mechanically jointed beans and columns.* Lecture B11 in Timber Engieering STEP 1, Centrum Hout, Alemere, 1995.

9 VTT Certificate No. 184/03. *Kerto LVL.* VTT Research Centre of Finland, Espoo, revised 24 March 2009.

10 Crews, K. *Development of high performance timber bridges in Australia.* Innovative Wooden Structures and Bridges, Lahti, Finland, IABSE, 2001.

11 *Specifying wood-based panels for structural use.* TRADA Technology, 2005.

12 *Plywood fact sheet.* No. 13. Wood for Good, 2005.

13 Canadian Plywood Association – Canadian Plywood – *Specifications Brochure for UK*, CANPLY, 2001.

14 Raknes, E. Durability of structural wood adhesives after 30 years ageing. *European Journal of Wood and Wood Products*, 55, 2–4, 1997.

15 Bengtsson, C. (ed.). *Glued in rods for timber structures,* Swedish National Testing & Research Institute, SMT4-CT97-2199, 2002.

16 *Sustainable timber sourcing.* TRADA Technology, 2007.

17 www.fsc.org (Accessed 10 April 2010).

18 www.pefc.co.uk (Accessed 10 April 2010).

5 Concept design

1 Bignotti, G. *Transportation and erection.* Timber Engineering STEP 2, Centrum Hout, Netherlands, 1995.

2 *Pre-fabricated modular timber bridges; Part 1. General description; Part 2. Manufacture of pre-fabricated parts and design selection; Part 3. Construction and launching; Part 4. Timber technology; Part 5. Typical design, 15m span four Truss bridge.* Restricted, UNIDO/IO/R.163, UNIDO, Vienna, 1985.

3 Fujiwara, T. *et al. Energy consumption through timber transportation and the woodmiles.* Third International Conference on Construction Materials, Vancouver, 2005.

4 The design and appearance of bridges. *Design Manual for Roads and Bridges*, Vol. 1, Section 3, BA 41/98, 1998.

5 *The appearance of bridges and other highway structures.* HMSO, 1996.

6 Cross-sections and headrooms. *Design Manual for Roads and Bridges*, Vol. 6, Section 1, TD 27/05, 2005.

7 Design criteria for footbridges. *Design Manual for Roads and Bridges*, Vol. 2, Section 2, BD 29/04, 2004.

8 The geometric design of pedestrian, cycle and equestrian routes. *Design Manual for Roads and Bridges*, Vol. 6, Section 3, TA 90/05, 2005.

9 *Passerelle de Vaires-sur-Marne (77).* CNDB, Focus chantier 15, Paris (accessed 12 October 2010).

10 Tommola, J., Jutila, A., Rautakorpi, H. and Wistbacka, J. *Design of wooden arch bridges.* Nordic Timber Council, 1996.

11 Solli, K. H. and Bjerke, H. *Timber truss bridges.* Nordic Timber Council, 1997.

12 Wells, M. *30 Bridges.* Laurence King, 2002.

6 Decks and parapets

1 Hislop, P. *Timber decking: the professionals' manual, 2nd. Edition.* DG2, TRADA Technology, 2006.
2 *Specifying timber species in marine and freshwater construction.* TRADA Technology, 2010.
3 Ritter, M. A. *Timber bridges: design, construction, inspection and maintenance.* US Department of Agriculture, Forest Service, EM 7700-8, 1990.
4 Taylor, R. J. and Keenan, F. J. *Wood highway bridges.* Canadian Wood Council, 1992.
5 Widmann, R. *Screw-laminated timber deck plates.* Innovative Wooden Structures and Bridges, Lahti, Finland, IABSE, 2001.
6 Crews, K. *Development of high performance timber bridges in Australia.* Innovative Wooden Structures and Bridges, Lahti, Finland, IABSE, 2001.
7 Crews, K. *Design procedures for stress laminated timber plate bridge decks: code and commentary.* University of Technology, Sydney, 1995.
8 Pousette, A. and Marklund, K.-A. *Stress-laminated bridge decks.* Nordic Timber Council, 1997.
9 Mäkipuro, R., Tommola, J., Salokangas, L. and Jutila, A. *Wood–concrete composite bridges.* Nordic Timber Council, 1996.
10 Abrahamsen, R. B. *Bridge across Rena River: "World's strongest timber bridge".* World Conference on Timber Engieering, Miyazaki, Japan, 2008.
11 Berthellemy, J. (ed.). *Les ponts en bois : comment assurer leur durabilité.* Sétra, 2006.
12 Pousette, A. *Wearing surfaces for timber bridges.* Nordic Timber Council, 1997.
13 Gustafsson, M., Nilsson, O. and Ström, T. *Railings for timber bridges.* Nordic Timber Council, 1997.
14 Gehri, E. (ed.). *Brücken und stege aus holz.* Schweizer bau dokumentation, Zürich, 1989.

7 Structural design

1 Gulvanessian, H., Calgaro, J.-A. and Holický, M. *Designers' guide to EN 1990. Eurocode: basis of structural design.* Thomas Telford, 2002.
2 *How to calculate the design values of loads using Eurocodes.* GD 2. TRADA Technology, 2008.
3 *Timber materials: properties and associated design procedures with Eurocodes.* GD 3. TRADA Technology, 2008.
4 Mettem, C. J. Timber structures, in Green, M. (Chair, T. G.) *Introduction to the fire safety engineering of structures.* The Institution of Structural Engineers, 2003.
5 Blass, H. J. *Columns.* STEP Lecture B6. Timber Engineering STEP 1, Centrum Hout, Netherlands, 1995.
6 Blass, H. J. *Buckling lengths.* STEP Lecture B7. Timber Engineering STEP 1, Centrum Hout, Netherlands, 1995.
7 Aune, P. *Shear and torsion.* STEP Lecture B4. Timber Engineering STEP 1, Centrum Hout, Netherlands, 1995.
8 Harris, R. J. L. (Chair, T. G.) *Manual for the design of timber building structures to Eurocode 5.* The Institution of Structural Engineers and TRADA Technology, 2007.
9 Tommola, J., Jutila, A., Rautakorpi, H. and Wistbacka, J. *Design of wooden arch bridges.* Nordic Timber Council, 1996.
10 Kreuzinger, H. *Dynamic design strategies for pedestrian and wind actions,* Technical University Munich, 2002.
11 *How to calculate deformations in timber structures using Eurocodes.* GD 5. TRADA Technology, 2006.
12 Bejtka, I. and Blass, H. J. *Joints with inclined screws.* CIB-W 18/35–7-5, Meeting 35, Kyoto, Japan, September 2002.
13 Bengtsson, C. and Johansson C.-J. (eds). *GIROD: glued in rods for timber structures.* Swedish National Testing and Research Institute Report 26, SMT4-CT97-2199, Lund, 2002.
14 Jorissen, A. J. M. *Double shear timber connections with dowel type fasteners.* Delft University Press, 1998.

8 Conservation, maintenance and repair

1 Tilly, G. *Conservation of bridges.* Routledge, 2002.
2 Pierce, P. C. *et al. Covered bridge manual.* Report No. FHWA-HRT-04-098. National Technical Information Service, Springfield, VA. USA, 2005.

3 *Timber bridge management.* Roads and Traffic Authority of NSW, Australia, 2002 (accessed 26 October. 2010).

4 www.rta.nsw.gov.au/cgi-bin/index.cgi?action=heritage.show&id=4300133 (accessed 8 November 2010).

5 www.timberbuilding.arch.utas.edu.au/projects/view_projectinfo.asp?projID=12 (accessed 8 November 2010).

6 Troive, S. *Environment and maintenance requirements on Swedish timber bridges.* Swedish Road Administration. NVF Conference on timber bridges. Hamar, Norway, 2005.

7 Highway structures: inspection and maintenance. *Highways Agency, design manual for roads and bridges.* BD 63/07, Vol. 3, Secn. 1. HMSO, 2007.

8 Ross, P. *Appraisal and repair of timber structures.* Thomas Telford, 2002.

9 Ritter, M. A. Bridge inspection for decay and other deterioration, in: *Timber bridges: design, construction, inspection, and maintenance.* US Department of Agriculture, Forest Service, EM 7700-8, 1990.

10 Makay, D. and Spivey, J. M. National Park Service, Historic American Engineering Record. *Addendunm to Contoocook Railroad Bridge.* HAER No. NH-38. Vermont, USA, 2003.

11 Brungraber, R. L. and Morse-Fortier, L. *Wooden peg tests: their behaviour and capacities as used in Town lattice trusses.* Vermont Department of Transportation and Massachusetts Institute of Technology, 1995.

12 Shanks, J., Walker, P. and Harris, R. *Development of rational guidelines for traditional joints in oak frame construction.* World Conference on Timber Engineering, 2008.

13 *Moisture in timber.* TRADA Technology, 2006.

14 *Non-destructive testing of timber.* TRADA Technology, 2001.

15 *Timber: fungi and insect pests.* TRADA Technology, 2004.

16 Bravery, A. F., Berry, R. W., Carey, J. K. and Cooper, D. E. *Recognising wood rot and insect damage in buildings.* 3rd. edition, CRC, 2003.

17 Ridout, B. *Timber decay in buildings.* Spon Press, 2001.

18 Dyken, T. *Monitoring of Norwegian bridges.* NVF Conference on timber bridges. Hamar, Norway, 2005.

19 Crews, K., Samali, B. Bakoss, S. and Champion, C. *Overview of assessing the load carrying capacity of timber bridges using dynamic methods.* Australian Road Bridges Forum, 2004. www.bridgeforum.org/files/pub/2004/austroads5/119_Champion%20AUSTROADS04.pdf (accessed 11 November 2010).

20 Fire protection COST Action C17 – Built Heritage: http://fred.english-heritage.org.uk/pdf/743.pdf (accessed 11 November 2010).

21 http://heritagefire.net/heritage_fire_objectives.html (accessed 11 November 2010).

22 Yeomans, D. *The repair of historic timber structures.* Thomas Telford, 2003.

23 Ritter, M. A. Bridge maintenance, rehabilitation, and replacement: case histories, in: *Timber bridges: design, construction, inspection, and maintenance.* US Department of Agriculture, Forest Service, EM 7700-8, 1990.

24 Commission for Architecture and the Built Environment. *UB 289 Moy Viaduct, Inverness.* www.betterpublicbuilding.org.uk/finalists/2004/moyviaduct (accessed 5 November 2010).

25 Mettem, C. J., Page, A. V. and Robinson, G. C. *Repair of structural timbers: Part 1 – tests on experimental beam repairs.* Research Report PIF 63, RR 1/93. Timber Research and Development Association, 1993.

26 Harris, R. J. L. (Chair, T. G.) *Manual for the design of timber building structures to Eurocode 5.* The Institution of Structural Engineers and TRADA Technology, 2007.

27 *Timber engineering hardware and connectors.* TRADA Technology, 2003.

28 *Glulam timber connectors.* Simpson Strong-Tie, 2010.

29 Racher, P. *Moment resisting connections.* STEP Lecture C16. Timber Engineering STEP 1, Centrum Hout, Netherlands, 1995.

30 Taylor, R. J. and Keenan, F. J. *Wood highway bridges.* Canadian Wood Council, 1992.

31 Mettem, C. J. and Milner, M. *Resin repairs to timber structures. Volume 1, guidance and selection.* TRADA Technology, 2000.

32 Resin-bonded repair systems for structural timber. TRADA Technology, 2001.

33 Broughton J. G. and Hutchinson A. R. *Review of relevant materials and their requirements for timber repair and restoration.* Oxford Brookes University, 2003. www.licons.org/documents/interimreports/08task2–2.pdf (accessed 6 November 2010).

34 Mays, G. C. and Hutchinson, A. R. *Adhesives in civil engineering*. Cambridge University Press, 1992.

35 Bengtsson, C. and Johansson C.-J. (eds) *GIROD: glued in rods for timber structures*. Swedish National Testing and Research Institute Report 26, SMT4-CT97-2199, Lund, Sweden, 2002.

36 Gustafsson, P. J. and Serrano, E. *Glued-in rods for timber structures: development of a calculation model*. Division of Structural Mechanics, Lund University, 2002.

37 LICONS: low intrusion conservation systems for timber structures. CRAF-1999-71216. www.licons.org (accessed 21 November 2010).

38 Gardelle, V. *Experimental database and first considerations on several low intrusive conservation systems for timber structures*. LRBB, Bordeaux, France, 2005. www.licons.org/minutes/Appendix%206%20-%20Task%20 5%20report%20-%20Vincent%20Gardelle.pdf.

39 Burgers, T. A., Gutkowski, R. M., Radford, D. W. and Balogh, J. *Composite repair of full-scale timber bridge chord members through the process of shear spiking*. Mountain Plains Universities Consortium, 2005.

40 Hutchinson, A. R. Surface preparation of parent materials, in: Zoghi, M. (ed.), *International Handbook of FRP Composites in Civil Engineering*, CRC, 2007.

9 General case studies and final recommendations

1 Lawrence, A. Modern timber bridges: an international perspective. *The Structural Engineer*, 16 September 2008.

2 *Le nouveau pont de Crest confirme le retour du bois dans les ouvrages de franchissement*. CNDB Dossier du Nois, Paris, France, January–February 2002.

3 Bogusch, W. *Holzbogenbrücke "Resgia."* Schweizer Holzbau, 9 September 1990. Lawrence, A. Modern timber bridges: an international perspective. *The Structural Engineer*, 16 September 2008.

4 Bridge for cyclists and pedestrians, Toijala. *PuuWoodHolzBois*, 4, 2001.

5 Puuinfo Oy, Finland, 2001.

6 Ollas overpass, Espoo. *PuuWoodHolzBois*, 4, 1998. Puuinfo Oy, Finland, 1998.

7 Ekeberg, P. K. and Søyland, K. Flisa Bridge, Norway: a record-breaking timber bridge. *Bridge Engineering*, 158, 1, 2005.

8 Miebach, F. Almere – Holzbrücke erobert Holland. *Mikado*, 10, 2008. www.mikado-online.de (accessed 30 November 2010).

9 *Passerelle d'Ajoux. Réalisations 2001*. www.cndb.org/?p=rechercher_des_realisations&table=xml_referentiel&fid=331 (accessed 26 November 2010).

10 Jäger, W. *Ausgewählte kapitel der tragwerkslehre – Holzerbrücke im forstbotanischen garten Tharandt*. Technische Universtat Dresden, 2007. http://tu-dresden.de/die_tu_dresden/fakultaeten/fakultaet_architektur/twp/studium/downloads/analysen/ss07/tharandt_mieth_nico.pdf (accessed 26 November 2010).